"So much information, so many years... This wonderful and much-needed book should be read by all 'horse people' for a more complete understanding of the role of the horse in not just China's history but in world history. Young's stunning illustrations validate the grandeur and nobility of horses as important contributors to Chinese culture. In addition to chronicling the history of horses in China, the book, importantly, details the current political, economic and human issues affecting China's horse population while suggesting solutions to continue and even expand the role of the horse in China. As a lifelong equestrian, I was intrigued by the horse's long and varied history described here. Not only enlightening but an enjoyable read, this book should be on the list for anyone who loves and works with horses."

Nancy Mazzoni
Lexington, Kentucky
Competitor and Thoroughbred Breeder

"An amazing and beautifully illustrated book. Not just about horse breeds, it also documents very well how horses have been cared for during China's history, and the great importance the horse once held in China, particularly in the era of the Song dynasty. Many of the drawings, pictures and scrolls, which all feature horses, are quite incredible. The mention of Marco Polo may surprise, as one only thinks of him as an explorer, interested in ships, not horses. An extraordinary amount of time and research has gone into this lovely book."

Richard Craddock
Founder, Craddock Advertising and Hyperion Promotions in Europe
Horse Breeder, New Zealand and Europe

"This is a wonderfully detailed and very readable book. It gave me considerable enjoyment to learn about the evolution of the horse in China. The very high quality of the book's illustrations greatly enhances the reader's experience in following the different aspects of China's long and varied equine history. I have great pleasure recommending *The Horses of China* to anyone seeking knowledge of a topic of which there is very little literature in the English language."

Roger Philpot
Former President
The Side Saddle Association, UK

"There is no shortage of books on horses, but few are devoted to Chinese hippology. *The Horses of China* is a comprehensive portrayal of Chinese equestrian life, celebrating the many different facets of this magnificent animal: the horse-human relationship, artefacts, sports, breeds and traditional Chinese veterinary medicine. Young does a marvellous job of collating fragmented yet fascinating material into a compelling and entertaining account. Her choice of illustrations is superb and makes the book a delight to explore. While it is interesting to understand Chinese culture from an equestrian perspective, Young also thoughtfully covers current horse issues. This well-researched and richly illustrated book is a source of real treasure for enthusiasts of the horse and Chinese culture."

Peter Yunghanns
Former Vice-President and Treasurer
The Federation of International Polo

The Horses of China

To Ruth and Alice

Young Yin Hung

THE HORSES OF CHINA

YIN HUNG YOUNG

Homa & Sekey Books
Paramus, New Jersey

FIRST EDITION

Copyright © 2021 by Yin Hung Young

Library of Congress Cataloging-in-Publication Data
Names: Young, Yin Hung, date- author.
Title: The horses of China / Yin Hung Young.
Description: First edition. | Paramus, New Jersey : Homa & Sekey Books,
   [2017] | Includes bibliographical references and index.
Identifiers: LCCN 2017046702 | ISBN 9781622460441 (hardcover)
Subjects: LCSH: Horses--History--China.
Classification: LCC SF284.C6 Y68 2017 | DDC 636.100951--dc23 LC record available at https://lccn.loc.gov/2017046702

Published by Homa & Sekey Books
3rd Floor, North Tower
Mack-Cali Center III
140 E. Ridgewood Ave.
Paramus, NJ 07652

Tel: 201-261-8810, 800-870-HOMA
Fax: 201-261-8890
Email: info@homabooks.com
Website: www.homabooks.com

Layout: Yin Hung Young

Printed in China
1 3 5 7 9 10 8 6 4 2

To all my horsey friends and
the horses I have met in my life

CONTENTS

ACKNOWLEDGEMENTS

I would like to express my sincere thanks to Nancy and Mike Mazzoni who a few years ago kindly invited me to their Thoroughbred farm in Kentucky. It was this eye-opening visit which set me on the path to researching material on the horse in China and to writing this book.

I am indebted to Carol Dyer of Word for Word in Hong Kong for the support and encouragement she provided me in pursuing the book's publication and for editing the manuscript.

I most gratefully acknowledge the tremendous assistance received from Guo Lianpo in my sourcing the book's images.

My thanks also go to Liu Ruikai who gave me advice on traditional Chinese equine veterinary medicine.

Finally, I extend my gratitude to GIMP (free image editing software) and the following parties for the permission given to reproduce photographs without charge:

- Emperor Qin Shi Huang's Mausoleum Site Museum
- Hong Kong Racing Museum
- Kerri-Jo Stewart
- Maoling Museum
- Metropolitan Museum of Art
- National Museum of China
- Osaka City Museum of Fine Arts
- Shaanxi History Museum
- Wang Zhanhong of CPC Liaoning Animal Domestic Genetic Resources Conservation and Utility Centre
- Wu Mingcheng and Xiang Mingyi of Heilongjiang Animal Reproduction Supervising Station and Northeast Agricultural University

Yin Hung Young

INTRODUCTION

When people talk of horses, no one ever thinks of China. China has never won any medals in major international equestrian events or produced any elite horses. Despite this, horses are very much a part of the history of China and have played an essential role in many different facets of the development of Chinese civilization.

In ancient China, horses were used in war, for transportation, for rituals and in agriculture. Horses, oxen, goats, pigs, dogs and roosters were considered the six major livestock animals, and of these horses were the most treasured. Known for their energetic and freedom-seeking nature, horses are also celebrated in the Chinese lunar calendar. Since their domestication in north-east China around 5,000 years ago, horses have had a profound impact on the destiny of the Middle Kingdom.

HORSES IN LITERATURE

In the *Shi Jing* (Book of Songs), which is the earliest existing collection of Chinese poems and songs, dating from as far back as 1000 B.C., horses are mentioned in 51 out of 305 poems, topping the animal list. The writings of Confucius are full of references to the horse; the horse makes a secular appearance in marriage, warfare, transportation, hunting and riding. The animal epitomized an intoxicating combination of national prowess, heroic bravery and fervour for life that provoked intense empathy in readers of the philosopher's work.

The *Zhuangzi* is a Taoism classic. It presents the horse's abstract and spiritual image in symbolizing freedom, ethereality and emancipation. Domestication leading to loss of instincts is seen as being against nature.

HORSE BREEDS

Native breeds which were first used by the Chinese as cavalry horses proved to be inferior to those against whom they fought. As conflicts with the northern nomadic Xiongnu became more pervasive, procurement of quality horses became a major priority. The imperial courts then developed their own breeding programmes, after which equestrian militarism changed war strategies, scales, complexities and finally the outcomes.

ART, SPORTS AND MEDICINE

The horse's influence over the civilization and art of China is unrivalled. A wide range of artefacts constructed on distinctive perceptual and cultural premises were created with horses as their motif.

Equine sports not only formed part of the education curriculum for young aristocrats, but also served as military preparation. As horses enjoyed a lofty status shared by no other animal at any level of Chinese society, equine veterinary skills became especially high and were a focus of traditional Chinese veterinary medicine, which was developed over 3,000 years ago and still serves as an alternative therapy today.

THE HORSE IN MODERN CHINA

Following the collapse of China's imperial dynasties in the early 20th century, the horse started a new journey and so did its riding partners.

Horses in China are mostly privately owned. While only a handful, which belong to the wealthy, lead a carefree life, the vast majority are kept perpetually busy. All over China, they plough the fields, pull carts, are used as pack horses, provide rides in tourist areas and are ridden for pleasure in riding clubs – from outskirts of big cities like Beijing, Shanghai and Guangzhou to their native habitats such as Xinjiang, Inner Mongolia, Qinghai and Yunnan. If they have no job to do and cannot prove their economic value, they are culled. The Chinese have a saying: "Working like an ox and a horse", which means working hard in a lowly position. However, work for the horse is getting harder to come by. With modernization of rural areas, pastoralists prefer to ride motorcycles to round up flocks of sheep, and ethnic minorities keep fewer horses than they used to in the light of the rising cost of living and government policies. It is no longer a common sight to see horses in the pastoral areas in the north-western and south-western border regions.

In the Changdu district of Tibet, more and more motorcycle dealers are springing up. Motorbikes are found parked in pastoral farms and by small dwellings all over the area. Before proper roads had been built, pastoralists relied on horses to get about. But horses can cover only 50 to 60 kilometres a day and it would take their owners a whole day to commute between other adjacent cities or counties. Now that roads and highways have been built, a motorbike with a full tank of petrol allows them to travel more than 300 kilometres a day. Yaks continue to graze the pasture land, but pastoralists on horseback are rarely seen nowadays. In 2011, the Tibetan government initiated a modernization plan whereby pastoralists can receive a 10 per cent subsidy for purchasing consumer electronics and motorbikes. Their modernized lifestyle is no more different from that of city dwellers, with households owning televisions, washing machines, smartphones and most importantly, motorbikes.

In Inner Mongolia, other than making their way to the annual Naadam Festival on horseback, Mongolians now primarily commute by motorbike in daily life.

LOSS OF GRAZING LAND

Land degradation, too, makes pasture land less available for grassland husbandry.

World Bank figures show that China has the third largest area of grassland after Australia and Russia. However, vast swathes of its grassland have become degraded; the Loess Plateau region and other extensive western areas have suffered especially badly. According to the People's Republic of China (PRC) National Bureau of Statistics, from 2004 to 2017, China lost 16.5 per cent or 43 million hectares of pasture land of which Inner Mongolia and Xinjiang accounted for 38 per cent and 36 per cent of the total loss, respectively.

Since 1996, the Chinese government has started to implement a nationwide grazing ban in selected areas with the aim of relieving overgrazing and erosion of the grasslands. This has inevitably had an adverse effect on the pastoralists' life. Few herders have the knowledge or the technological hardware to provide for their animals in feedlots, and the cost of livestock rearing is consequently rising. Mongolian horses, in particular, have been kept in the open since the dawn of history; being fed in sheds is against the age-old practices of their owners — and indeed the instinct of the animals themselves. With a lower economic value than other livestock, such as beef cattle, goats and sheep, horses are no longer the pastoralists' first priority. They keep fewer livestock than previously and horses are first among those they get rid of.

In mid-2016, the Inner Mongolia government reported that desertification had moderated, with deserts in the province being reduced from 2009 by some 416,900 hectares to 60.92 million hectares, which is about 51.5 per cent of the province's land. Meanwhile, desertified land, defined as that which is on the brink of being turned into desert, was decreased by 343,300 hectares to 40.78 million hectares.

AND OF THE FUTURE?

Through showcasing the chronicled history of China's horses, equine art, equine sports, horse breeds and traditional Chinese veterinary medicine, this book reveals the philosophy, value system and wisdom of Chinese civilization. The role of the horse is expounded upon in this journey, for the world's fastest distance-running quadruped has undoubtedly helped to build the country historically.

The question now is: how will China help to build the future for this steadfast animal which has served it so well, and care for its welfare?

CHAPTER ONE

HUMANS AND HORSES: ANCIENT CHINA TO THE PRESENT DAY

INTRODUCTION

Chinese civilization is widely regarded as being the fountainhead of all Eastern culture. It can claim four great inventions which gradually spread throughout the world and wielded an enormous global impact: gunpowder, paper making, printing and the compass. Nevertheless, its development of an equestrian culture lagged far behind that of other civilizations. The relationship between horse and man was a culture which was imported into China, but once localized it flourished throughout the empire.

ORIGINS OF EQUESTRIAN CULTURE IN CHINA

The *Shuowen Jiezi*, the first Chinese dictionary written between 100 and 121 in the Eastern Han dynasty, contains 115 characters with the horse as a radical covering horse types, horse motions, horse personalities, equestrian gear, official titles, state names and human names. The number and scope of these characters reflect the rich, deep and wide influence the horse had at that time and also an understanding of the animal's use. As the dynamics continued, the role of the horse changed and equestrian lore grew. In the *Kangxi Dictionary*, the standard dictionary used during the 18th and 19th centuries, the number of horse-related characters increased to 658.

The military value of horses was highly recognized. The authorities in ancient China did not develop any policies for animals except for horses. Equine policies and officials were of tremendous significance to the empire. A strange story is told of the famous monkey king. The Jade Emperor offered him the position of Protector of the Horses to watch over the celestial mews; it was a senior position yet the monkey was told it was the lowest. Misconstruing the Jade Emperor's intention as an insult, the furious monkey wreaked havoc in heaven. As two other examples of the horse's significance, an ancestor of the Qin dynasty rose to prominence by successfully breeding military steeds for the king; and An Lushan, who initiated the catastrophic An Shi Rebellion decisively weakening the Tang dynasty, had been given extensive jurisdiction over the empire's cavalry horses.

Throughout the long and intricate history of the Middle Kingdom's settled civilization, no other animal has made an impact comparable to that of the world's fastest distance-running quadruped on its destiny. From diplomatic policies to the militaristic consolidation of territory, the horse has always played its part.

EARLY DOMESTICATION

In all likelihood, the horse was first domesticated about 6,000 years ago somewhere on the Eurasian Steppe; this was approximately 1,000 years earlier than its appearance in China.

China's first horse stock may have originated from a domesticable variant of the wild Przewalski's horse. After being crossed with some imported stock, undoubtedly through selective breeding, these horses had by the 10th century B.C. evolved to appear like the modern, pony-sized Mongolian horse.

HORSE-DRAWN VEHICLES

In most ancient civilizations, carriages were first pulled by oxen and only later were drawn by horses.

The chariot first emerged in China around 1300 B.C., some 1,400 years after its invention in the Fertile Crescent.

The earliest definite evidence of the horse's use in wheeled transport in China was found in the funeral site of King Wuding at Anyang in Hebei (Fig. 1.1) and dates back to between 1250 and 1192 B.C. during the Shang dynasty. These conveyances were all single-shaft chariots (Fig. 1.2) which required at least two horses to haul.

The use of chariots as a means of transportation at this time is in no doubt, but the practical military effectiveness of the horse-drawn conveyance is suspect. Historical finds show that the two-horse chariot of the Shang dynasty and the four-horse chariot of the Western Zhou dynasty (1046–771 B.C.) were very heavy and difficult to manoeuvre; they could only be used efficiently on relatively flat and smooth terrain. The *Sunzi Bingfa* (Art of War) suggests that the chariot was used more as a mobile command post rather than as an actual fighting machine. Despite this,

Fig. 1.1 The earliest definite evidence of the horse's use in wheeled transport in China was found in the funeral site of King Wuding at Anyang in Hebei and dates back to between 1250 and 1192 B.C. in the Shang dynasty.

Fig. 1.2 A single-shaft carriage in the Spring and Autumn period (770–476 B.C.). Such conveyances required two or more horses to pull them. (Reference: Liu, Y. H. (2013). *Carriages and Harness of Ancient China* (p. 59). Beijing: Tsinghua University Press.)

chariots still had a role as a status symbol. In those times, a state's prestige was measured by the number of war chariots and horses it possessed, and the burial of horses and chariots alongside someone of renown was common among the privileged and the aristocracy.

By the late Eastern Zhou dynasty (770–256 B.C.), chariots had been replaced by intricately decorated individual weapons as prestige items. The Xiongnu, a nomadic people from North Asia who controlled a vast area stretching from Manchuria in the east to Turkestan in the west, posed a sinister threat to the Chinese over a long period. With the increasingly frequent raids by the Xiongnu along the northern border and the rise in the importance of cavalry, the role of the chariot became significantly diminished.

RISE OF MOUNTED CAVALRY

Riding on horseback was viewed as barbaric in ancient cultures, while riding behind horses was considered civilized.

During the Warring States period (475–221 B.C.), the state of Zhao had long been harassed by neighbouring nomads and other states. King Wu Ling concluded that the costumes, cavalry and archery of the nomads were all better and more functional than the dress, chariots and spears used by the Chinese on the battlefield. Determined to improve the prowess of his army, he decreed in 307 B.C. that Zhao soldiers should wear nomadic costumes and practise horse riding and archery. So determined was he that he acted as an exemplary model in demonstrating these orders himself.

King Wu Ling's revolutionary reforms not only lessened the Chinese traditional disdain for nomadic customs but also led to the rise of the cavalry in Chinese military history.

At that time, the state of Qin was the only kingdom comparable to the state of Zhao in terms of equine military forces, both possessing 10,000 troopers. The Qin people lived in the western border areas and had fought and mingled with nomadic tribes for generations. They were also regarded as barbarians and ostracized by central kingdoms. Nevertheless, the geographical advantages bestowed them superior equestrian skills and lore, establishing a solid foundation for them to unify China.

HORSES IN CHINA'S IMPERIAL HISTORY

QIN DYNASTY (221–206 B.C.)

Once the conquest of the other six states had been completed in 221 B.C., the Qin established the first unified, multinational and power-centralized state in Chinese history.

According to the *Zhanguoce* (Intrigues of the Warring States), the Qin army included 1 million armoured warriors, 1,000 chariots and 10,000 troopers. As of today, more than 7,000 terracotta soldiers and 600 horses have been excavated from the Qin Shi Huang mausoleum. Based on the suggested ratios, the cavalry was just a minor part of the military force and was there mainly to supplement the infantry and the chariots.

Back in the Qin dynasty however a governing body in charge of horses had already been set up to oversee equestrian issues. Its responsibility ranged from horse breeding and maintenance of royal mews, harnesses and horse-drawn carriages to the training of grooms, troopers and carriage drivers. The governing body's influence continued throughout the feudal period.

During the Qin dynasty, progressive legislation was brought into force to ensure the quality and quantity of cavalry horses.

An annual examination of official horses was held across the country. If a horse was found to be emaciated, slow to learn or missed any trial examination, the official in charge would be fined with an amount equivalent to the cost of a shield. During military service, if a horse received a low grade on evaluation, the groom in charge would pay a fine equivalent to the value of a coat of armour; other responsible officers would each pay a fine equivalent to the value of a shield. The fine for hurting a carriage horse depended on the length of the cuts, abrasions or scratches found. If these measured the equivalent of 2.3 cm, the fine would be that of the value of a shield; if 4.6 cm, the fine equated to two shields; if the wound was more than 4.6 cm, the fine was one coat of armour.

Geldings had to be kept separately behind carriage horses and could not be whipped. A carriage horse's harness was to be removed once the horse had finished its work; if this was not adhered to, the fine would be equivalent to the value of a shield.

The Jueti was a premium horse breed in northern China in ancient times. A trainer had to train up at least six Jueti every year. If he failed to do so, he had to pay a fine of an amount equal to a shield.

If a cavalry horse did not behave and listen to instructions, the officials in charge were fined with an amount equal to one or two coats of armour.

Every head of cattle or horse had to bear a marking for identification purposes. If the marking was wrong, the official groom or farmer was subject to a fine equal to the value of a shield.

These fines were considered heavy, especially by local standards. Anyone who could not afford to pay had to perform corvée (bonded labour) to pay off his debt. Using this system, the Qin always had sufficient military supplies in reserve and plenty of available statutory labour.

HAN DYNASTY (206 B.C.–A.D. 220)

The Han dynasty is commonly considered a high point in the history of imperial China.

During the Han times, China experienced an immense change in the equine population.

At the beginning of the Han dynasty, only a limited number of war horses were left alive after the battles that had brought the dynasty to power. It was hard to find four steeds of the same colour to draw Emperor Gaozu's carriages, and senior officials, and even generals, had to rely on ox-drawn carriages.

As cavalry had replaced the chariot as the major force in use in warfare, raising an ample supply of war horses and procuring superior stock for defence against the Xiongnu became of prime importance.

In the early Han dynasty, a governing body in charge of war horse husbandry was set up at both central and district levels. Empress Lu Zhi banned the export of mares to prevent the military's supply from falling into the hands of their enemies. During the time of Emperor Wendi and Emperor Wudi, if a family kept a horse, three family members would be excused performing corvée. Emperor Jingdi established 36 stud farms

Fig. 1.3 Introduced in the Eastern Han dynasty (25–220), the double-shaft carriage could be pulled by only one horse. This plate also illustrates the breast-strap harness, invented c. 400 B.C., which helped maximize the horse's hauling power without hampering its ability to breathe. (Reference: Liu, Y. H. (2013). *Carriages and Harness of Ancient China* (p. 200). Beijing: Tsinghua University Press.)

Fig. 1.4 Zhang Qian's diplomatic expedition to territories to China's west is depicted in the Mogao Caves in Dunhuang, Gansu. The upper right of the mural shows Wudi and his officials worshipping a golden idol at the Sweet Spring Palace captured from the Xiongnu by Huo Qubing in 121 B.C. In the bottom middle of the mural, the emperor is seen riding a horse, followed by his officials. With a retinue behind, Zhang Qian holds a ritual tablet and kneels down to the emperor as he bids farewell to him in 138 B.C. before his diplomatic quest starts. Zhang Qian's actual journey is represented in the middle left of the mural. On the upper left is a castle in Daxia, in what is now northern Afghanistan, and a destination of the expedition. Although the events depicted may not be in true chronological order, the mural provides an historically important equestrian record.

and those with money. To make the most of available equine resources, the double-shaft carriage (Fig. 1.3), which needed only one horse to pull it, was invented to replace the single-shaft carriage, which required two.

On a separate note, in the Eastern Han dynasty (25–220), the design of the saddle was improved with a higher pommel and cantle. This increased the stability of the rider's seat and heralded the mass adoption of cavalry.

THE HEAVENLY HORSES

Emperor Wudi (reigned 140–87 B.C.) of the Han dynasty was among the greatest of all Chinese emperors. He came to power as a young boy, but the precocious teenage emperor made his reign not only one of the most prosperous periods in Chinese antiquity, but he also created one of the most powerful regimes in the world.

As military clashes with the Xiongnu turned more belligerent, he sent Zhang Qian (Fig. 1.4) in 138 B.C. to form alliances with western states hostile to the Xiongnu. The attempt at diplomacy suffered a setback and Zhang Qian was captured by the Xiongnu and held for 12 years before being released in 126 B.C. In Fergana (modern day Uzbekistan), Zhang was to encounter the marvellous horses which during exertion would discharge small amounts of blood mixed with sweat to form a pink-coloured foam.

in north-western provinces. The Han also obtained horses through trading with the Xiongnu. As a result, they had 300,000 mounts at their disposal in defence against the Xiongnu. According to the *History of Horse Rearing in China* by Xie Cheng Xia, such large-scale horse rearing was rare even in Western civilizations at that time.

As Han society became wealthier, the demand for carriages increased among aristocrats, officials

The Chinese later called these animals "blood-sweating purebreds" or "heavenly horses".

These horses were thought to be Nisaean horses originating from the Nisaea plain where the Persian royal stud farms were located. The clover that grew in the plain was known as Median grass, recognized today as alfalfa or lucerne, which is a highly nutritious legume fodder with twice as much protein as most grass hays. Excellent forage plus

Fig. 1.5 Yili or Wusun horses graze under the Mountains of Heaven, which straddle the border between China and Kyrgyzstan

selective breeding enabled Nisaean horses to reach their fullest genetic potential.

Emperor Wudi was captivated by the blood-sweating purebred described by Zhang, both because of China's military need for superior cavalry stock and his belief that the horses were the legendary heavenly horses which could bring him immortality. According to the *Classic of Mountains and Rivers*, riders of the celestial creature, which has a white body, a black head and a pair of wings, could live for up to 2,000 years. The Fergana Range is known as the Tian Shan (Mountains of Heaven) in Chinese (Fig. 1.5). Adding this to his imagination, the emperor likely deduced that the Nisaean was the legendary heavenly horse.

Han Wudi launched the first of several military quests for these coveted "heavenly horses" in 104 B.C.

However, when Fergana not only refused to sell their prize horses but also murdered the Han envoys, Wudi was infuriated, ordering Commander Li Guangli to wage war on Fergana in retaliation and to seize the horses by force. Three years later, after spending staggering sums on the campaigns and losing some 70,000 men, who died of starvation, exposure and exhaustion in the rigorous march or were killed on the battlefield, the Han army finally defeated Fergana and was able to take home dozens of Nisaean horses and 3,000 of other breeds, many of them Akhal-Tekes, both mares and stallions. However, by the time they reached the capital Chang'an, only about 1,000 horses had survived the arduous journey. In other words, it took the lives of some 70 soldiers to obtain a single horse.

Horses were equally precious in the ancient Near East. In Assyria, for example, a horse was equivalent in value to 30 slaves or 500 sheep.

Since time immemorial, every country has attempted to prevent potential adversaries from acquiring any of its military assets that could then be used against them. While arms embargoes are a commonly seen current form of sanctions, back in the Middle Ages, Spain, France and England all imposed a ban on the exporting of horses. England under the Tudors even prevented the export of horses to Scotland.

In 121 B.C., Wudi had already sent envoys to Parthia to look for Nisaean horses and to form an alliance against the Xiongnu. Mithradates II was king of the empire at that time and was considered the greatest of the Parthian kings. It was under Mithradates II that the empire reached its grandest size and glory and that official trade relations via the Silk Road had developed.

THE WUSUN HORSES

Another major horse expedition was led by Zhang Qian in 115 B.C. when he was sent on the second diplomatic mission by Emperor Wudi. In the Yili River Valley basin, he happened upon a further nomadic tribe, the Wusun, who possessed horses which although neither as strong nor as large as those from Fergana were definitely superior to the native Chinese stock. These were thought to be a hybrid between the horses of Fergana and the Mongolian-type horse. As a result of Zhang Qian's embassy, the Han court received a large number of Wusun horses in two rounds of tribute.

THE CAVALRY

Even before the acquisition of the heavenly horses and the Wusun horses, the Han had built up a mighty cavalry force, which proved powerful enough to keep the Xiongnu at bay and prevent them from making increasing incursions on the northern frontier.

By then, China had already been mired in civil war from the fall of the Qin dynasty through the Chu-Han Contention, during when the Xiongnu had grown stronger and gradually occupied territory in China's north-west. To defend the empire and to expand its territory, in 200 B.C., Emperor Gaozu led a military force of 320,000 men to subjugate the Xiongnu, chasing them as far as Pingcheng (present-day Datong, Shanxi). Underestimating the strength of the Xiongnu ruler, Modu Chanyu, Gaozu was then ambushed by Xiongnu cavalry and was besieged in Baideng for seven days. The Han soldiers were unaccustomed to the harsh winter in the north and many suffered from frostbite and hypothermia. Food supplies were soon exhausted and the spirit of the Han plunged. Finally, Gaozu followed his advisor Chen Ping's suggestion to bribe Modu Chanyu's wife, who in return successfully persuaded Modu to retreat.

Emperor Gaozu realized from this setback that the Han were no match for the Xiongnu. In the ensuing 60 years, the Han launched a passive foreign policy by forging royal marriage alliances with the Xiongnu and paying them great amounts of tribute in the form of silk clothes, food delicacies, wine, exquisite handicrafts and intricate gold ornaments.

Fig. 1.6 The horse collar, in use by c. 100 B.C., distributed the load around a horse's neck and shoulders when hauling a wagon or plow. (Reference: Liu, Y. H. (2013). *Carriages and Harness of Ancient China* (p. 260). Beijing: Tsinghua University Press.)

The Xiongnu however did not comply with the agreement drawn up on the cessation of hostilities; skirmishes and military conflicts continued, and they pursued their invasion of the north-west. Tensions then rose to new heights under Wudi.

By the time Wudi had ascended to the throne, not only had the Han empire recovered and accumulated abundant wealth, but the equine and human populations had also both grown rapidly. The Han were now well equipped and capable of counter-attacking the Xiongnu. From 133 to 119 B.C., Wudi initiated several military campaigns against the Xiongnu in the Ordos Loop, the Hexi Corridor and the Gobi Desert. The Han achieved a victory in most of their campaigns, notably in the Battle of Mobei.

In 119 B.C., Hui Qing and Huo Qubing each separately led 50,000 troopers and several hundred thousand infantrymen to ram the Xiongnu's established base in the Gobi Desert in a pincer movement. This prodigious Han expedition ended up decimating the Xiongnu military force and overrunning its rear. The Xiongnu completely fell apart and deserted en masse. The Han claimed the territory.

The Han empire's narrow but decisive triumph brought about long-term peace along its north-western borders. The Xiongnu were all but wiped out from the southern Gobi Desert and driven to the barren north. Some scholars believe that they later fled to Central Asia and then on westwards towards Europe. Here they were called Huns and became a grave threat, not least in their devastating incursions into the Balkans, Thrace and the Eastern Roman Empire during the reign of Attila in 445–453.

From the time of Wudi right up until the end of the Han dynasty, the Xiongnu remained unable to regain the territory they had lost in the Middle Kingdom. Large-scale encroachment by the Xiongnu only took place again some 400 years later during the Jin dynasty.

The breast-strap harness, the horse collar and the stirrup are China's most significant contributions to the equestrian culture of the world, each advance helping realize some tangible increase in the horse's capability and redefining the use of the horse across different sectors.

□ Breast-strap harness (c. 400 B.C.). This was the first harnessing system which effectively used the horse's pulling power without restricting its breathing (Fig. 1.3). The shafted horse-drawn vehicle was developed on the foundations of this invention; it was heavy but capable of carrying up to six passengers drawn by only one horse. In contrast, much lighter vehicles pulled by four horses with a maximum load of two passengers were in use by other ancient civilizations in the same period.

□ Horse collar (c. 100 B.C.). The most efficient harness in ancient China, the horse collar transformed a horse into an ox-substitute by creating a collar and an artificial (ox-like) "hump" at the top of its neck (Fig. 1.6). It allowed the horse to use all its strength when hauling, in essence enabling it to push forwards with its powerful hindquarters into the collar. The collar was padded to avoid discomfort and to protect the horse's shoulders.

□ Stirrups (c. 300). Stirrups were first used in China, although some Chinese experts nevertheless speculate that it was the nomad, the Xiongnu or Xianbei, who invented the stirrup.

In a tomb dating back to 302 in Changsha, Hunan, excavations were made of several tacked horse figurines. Each features a triangle lying to the front left of the saddle. This is the earliest depiction of a stirrup, but at that time, it was solely used for mounting purposes (Fig. 1.7).

A single stirrup which formed part of a full tack set was excavated in a tomb at Xiaowen Tun, Anyang, Henan, dating back to the late Western Jin to the early Eastern Jin (c. 300). The stirrup is 27 cm high with a long shaft and an oval frame of an outer diameter of 16.4 cm and a width of 1.8 cm (Fig. 1.8).

In the Yuan Tai Zi Tomb, Chaoyang, Liaoning, which dates back to the Eastern Jin dynasty, a pair of stirrups was dug up together with the rest of a complete set of tack. These stirrups were made of wood, covered with leather and finished with paint in a cloud pattern (Fig. 1.9).

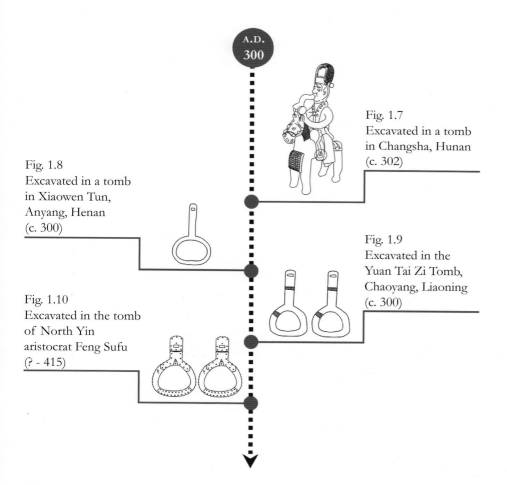

A.D. 300

Fig. 1.7
Excavated in a tomb in Changsha, Hunan (c. 302)

Fig. 1.8
Excavated in a tomb in Xiaowen Tun, Anyang, Henan (c. 300)

Fig. 1.9
Excavated in the Yuan Tai Zi Tomb, Chaoyang, Liaoning (c. 300)

Fig. 1.10
Excavated in the tomb of North Yin aristocrat Feng Sufu (? - 415)

Figs. 1.7–1.10 These excavations indicate that the stirrup was in use in China as early as A.D. 300.

The stirrups in the tomb of a North Yin aristocrat, Feng Sufu (?–415), Liaoning, feature a wooden triangular framework coated with gilded bronze sheets and a short shaft. They are 23 cm high by 16.8 cm wide (Fig. 1.10).

The use of stirrups enabled riders to mount more easily, balance themselves and far better control their horses. Stirrups also provided them with a secure platform from which to fight more effectively with weapons at close quarters.

The invention was made using the metal casting technique, which had already been mastered by the Chinese over a millennium earlier.

The benefit of stirrups was soon appreciated and their use spread eastwards to Japan and Korea, and westwards during the Middle Ages through the nomadic peoples of Central Eurasia to Europe.

Joseph Needham, a British scientist, historian and sinologist, considered the impact of stirrups was as significant as that of gunpowder in Europe. In his opinion, although they were very simple, stirrups helped the advancement of horse tack and the development of cavalry, thus setting up and solidifying feudalism in Europe, which however was finally shattered by gunpowder – also invented by the Chinese.

HORSES AND THE SILK TRADE

As early as the Qin dynasty, silk served as a currency for trade with the nomads in exchange for horses. Silk for horses and later tea for horses became more common throughout imperial history. As the Chinese became increasingly desperate for horses, the bargaining power of the nomads in this trading gradually grew.

THREE KINGDOMS PERIOD, JIN DYNASTY, SOUTHERN AND NORTHERN DYNASTIES (220–589)

This was a period of chaotic warfare, yet with significant artistic, cultural and technological advances, in particular during the Southern and Northern dynasties when nomadic tribes invaded China and took power. Northern states were of pastoral origin and had military prowess, fielding heavy cavalry more often than the Southern states. The cavalry of this period marked an epoch,

rallying hundreds of thousands of troopers in military campaigns. In 383, Fu Jian of the Former Qin mobilized 270,000 troopers to combat the Eastern Jin, which had only 80,000 troopers, in the Battle of the Fei River. Fu Jian bragged about his massive cavalry force, saying, "If the troopers all dropped their whips into the Yangtze River, its flow would be stopped." Emperor Taiwu (reigned 423–452) of the Northern Wei sent 600,000 troopers to cross the Huai River, invading the Liu Song dynasty territory in the south.

The mass adoption of cavalry, notably heavy cavalry, occurred in the wake of the invention of stirrups, the advancement of metallurgy, the migration of nomadic tribes and the improvement of horse breeds.

Advanced metallurgy made it possible to craft a large quantity of plated-mail armour and stirrups. Stirrups helped troopers to enhance their mobility, stability and striking power, to sustain long marches and to fight in close combat.

The steppe tribesmen brought with them to the Middle Kingdom their long-standing equestrian traditions, sophisticated cattle husbandry and superior mounts. These animals could breed stronger and more enduring horses that could carry the weight of the armour (Fig. 1.11).

TANG DYNASTY (618–907)

The Tang dynasty is regarded by historians as the most glittering period of Chinese civilization – surpassing the earlier Han dynasty.

China's second massive change in the equine population occurred in the Tang dynasty.

The royal family of the Tang court was descended from the military aristocracy of the north-west and had intermingled with steppe nomads over several generations. During the overthrow of the Sui dynasty, Emperor Gaozu bought 2,000 horses and borrowed troopers from the Turkic peoples. While the Sui army predominantly used heavy cavalry as its striking force, the Tang, influenced by the Turkic peoples, fought with light cavalry whose chargers were armour free. This is evidenced from the famous stone relief panels, *Six Steeds of Zhaoling Mausoleum* (see the Horse in

Fig. 1.11 A pictorial brick of the Southern dynasty (420–589), excavated in Henan, depicts two Nisaean horses, one of which wears armour similar to that worn by Parthian cataphracts.

Art and Equestrian Artefacts chapter). With their higher mobility and stamina, the Tang cavalry defeated the sheer offensive power of the Sui cavalry. In the early Tang, the cavalry was its core military force; the ratio of troopers to infantry in battle was three to one.

Against the backdrop of the Tang court's part pastoral origin and the significance of its cavalry, it goes without saying that its horses were highly treasured.

The Tang cavalry was the domain of the elite. They provided their own horses and weapons; and membership therefore was prohibitive to all but the wealthy.

A mail system connected by horses dates back to the Qin dynasty, when a somewhat unsophisticated but nationwide network had already been established (Fig. 1.12). The system became more advanced in the Tang when a court message could be conveyed over 200 kilometres in a day. Some 1,639 postal relay stations were established across the country, and of these 1,310 were connected by horses, 260 were connected by boats and the remaining 69 used a combination of both. They were maintained at intervals of 50 kilometres along the main routes crossing the country. Each equine station covered four hectares of land and kept eight to 75 horses, depending on the importance of the station. The relay horses were well treated and had one groom appointed to look after every three of them.

Emperor Tang Xuanzong's favourite imperial concubine, Yang, was crazy about lychees, a delicious, juicy red-skinned fruit native to southern China. Through the express system, the plump beauty could always enjoy fresh lychees sourced in summer from Fuling, Chongqing, which was over 700 kilometres away from the capital, Chang'an.

Laws were enacted by the Tang to protect the interests of both the government and the horses. Stealing and then slaughtering a government horse was the most serious offence and could lead to a sentence of two and a half years of hard labour. Overloading a postal relay horse, riding a postal relay horse off the highway and passing a postal station without changing horses were all punishable offences.

Through concerted effort, horse-breeding programmes reached a pinnacle in Tang times. The common Mongolian-type horses had been interbred with a variety of bloodstock from the

Fig. 1.12 A pictorial brick of the Southern and Northern dynasties which features a mounted postman without a mouth, implying confidentiality of the message.

5,000 in 618 to 700,000 by 668. The programme not only met the dynasty's appetite for horses but also improved the stock quality.

The aristocrats' great love of horses was likewise reflected in leisure and recreational activities, and was instrumental in horse dancing, polo and horseback hunting reaching an apex at this point in China's imperial history.

However, all the flourish and feats of accomplishment collapsed in the final years of the Tang rule.

SONG DYNASTY (960–1279)

Unlike the Tang, the Song did not have an equestrian or militaristic heritage. Infantry was the Song's major military power; the scale and strength of its cavalry were greatly inferior to the Tang's.

During the Song period, north-east China was occupied by the ethnic Western Xia empire, while the north-west was occupied by the Liao and later by the Jin; the horse breeding base was thus lost.

Tibetan horses, through the "tea for horses" market, were a major source of horses for the Song, but they were adapted to high altitude plateaus with thin air, low atmospheric pressure and freezing weather. Living in the China plains, in humid and warm weather, caused the horses serious health issues, especially in their pulmonary systems. In battles against the Liao empire in the northern plain, Tibetan horses were unable to reach the fullness of their true potential.

Given the shortage of chargers, the Song cavalry represented only one-seventh of the military force. Even worse, 30 to 40 per cent of the Song troopers did not have chargers and in extreme situations up to 90 per cent of them were unmounted. The Song horses were tiny, standing only 132 cm or 13 hands high while each nomadic trooper had at least one reserved charger that was bigger than the Song's.

Central Asian states of Kokand, Samarkand, Bukhara, Kish, Chack, Maimargh, Khuttal, Gandhara, Khotan and Kyrgyz. Horse trade was robust, especially in the latter half of the dynasty, to ensure a premium horse supply. Tibet sold horses to China in exchange for both tea and silk, while the Uyghur traded only for silk.

In 651, following the invasion of Iran by Muslim Arabs, Yazdgerd III, the last king of the Sassanian Empire, was captured. His two sons, Peroz III and Bahram VII, together with most of the surviving Persian nobles, fled to China, bringing along many imperial horses. Peroz was given the title of "Awe-inspiring General of the Left [Flank] Guards".

Sino-Arab diplomatic relations formally started in the same year when the Arab Islam Caliph Uthman Ibn Affan (reigned 644–656) sent an envoy to the Tang court. Although the envoy refused to pay homage to Tang Xuanzong as Muslims bowed only to their god but never to a prince, he presented several premium Arab horses to the royal court.

The Tang court also concocted an intricate structure for tending its herds and employed strict laws governing the treatment and handling of the royal steeds. The stock of horses had grown from

The serious horse shortages faced by the Song partially contributed to the dynasty's overthrow by the Mongols.

YUAN DYNASTY (1279–1368)

China's salvation had rested with the fragmentation of the nomadic tribes until 1211, when the great Genghis Khan (1162–1227) unified these Mongol tribes into one sweeping confederacy which then convulsed the world. The intrepid Mongol warriors were by far the most effective and ruthless cavalry forces of all time. They were said to be able to travel for 10 days, getting by on milk from their mares, which made up the majority of their chargers, and in the worst situations, horse's blood, which they drank from a pierced vein.

The Mongols' success in sweeping through Central Asia and Europe, despite their being often vastly outnumbered by those they met in their path, was the result of their cavalry tactics and the advantages they had developed over centuries.

The formidable and infamous Mongol armies did not directly confront their rivals with sheer numerical superiority, but would send mounted detachments to create chaos, confusion and panic through skirmishes and by attacking their rivals' flanks and rear. If this approach succeeded in causing utter disorganization among their opponents, they would finish them off by raining lances and arrows at them.

Facing cataphracts, a form of heavy cavalry in western Eurasia and the Eurasian Steppe where horse and rider wore whole-body armour, the elusive Mongol troopers and horses were clad in only lightweight leather amour that gave them the advantage of unequalled mobility to scout, spy, escape, ambush, sneak and strike. Mounted Mongol scouts could travel at incomparable speeds in a day to share intelligence. Each Mongol trooper typically kept three or four horses in reserve to prevent the horses from becoming exhausted while at the same time maintaining high mobility. Such mounted spies would find campsites with plentiful natural resources so that both soldiers and horses could live off the land. This reduced the logistical burden of carrying supplies.

Other siege tactics included feigned retreats, unexpected attacks, hostage taking, psychological warfare and the use of human shields. The Mongols also hired intelligent Chinese and Middle Eastern engineers to produce highly destructive catapults and other siege devices.

Despite all this, their traditional military advantages and cavalry tactics could not be fully deployed in their conquest of the Song dynasty. The Mongol Empire took more than 40 years to subjugate the Song but needed far less time to defeat other countries: they took Siberia in one year (1207), Qara Khitai (1216–1218) and Khwarazm (1219–1220) in two years each, the Western Xia empire in 21 years (1205–1227) and the Jin empire in 22 years (1211–1234).

From 1235–1278, the Mongols launched three large-scale military campaigns against the Song. Their early raids on the Song were not conducted by their elite forces, who were at that time engaged in plundering and sacking communities across Europe. The Middle Kingdom, unlike Europe and Central Asia, did not have the same vast regions of grassy plains which played into the hands of the masters of the steppe. China nevertheless had a booming economy, fuelled by a prodigious number of exports to Central Asia, Southeast Asia, Korea, Japan and even as far as Africa, ranging from silk, tea and porcelains to printed books, paper and copper coins. In addition to its strong agricultural base, the Song therefore had plenty to rely on to sustain a long-term battle. The Mongols favoured winter for their campaigns as the season had the climate closest to that of their inhospitable homeland upon which many of their ground tactics had been developed. However, although they might open their military campaigns against the Chinese Song in winter, the campaigns often dragged on through the summer when the Middle Kingdom was so sultry that neither the Mongolian horses nor the men could cope with the heat. Worse still, the conditions encouraged epidemics to spread in the Mongol camps.

The Mongols came particularly unstuck by the major and critical battles that took place in Sichuan and Xiangyang, Hunan, where naval rather than cavalry forces and tactics were most needed.

This was also their Achilles heel that led to their failed expeditions against Vietnam and Japan.

Helped by the recruitment of Pu Shougeng, a Muslim defector who had been in charge of the Song's maritime affairs for 30 years, the situation changed. Pu Shougeng handed over all the Song's warships to the Mongols and helped to build another 600. With this addition of naval hardware, the Mongols made great tactical strides. Finally, exacerbated by internal political conflict, the Song was toppled by Genghis Khan's grandson Kublai (1215–1294) who proclaimed the Yuan dynasty in 1279. The Yuan was the first alien dynasty in Chinese antiquity.

Overwhelming destruction of the royal herds was among the first acts of the Mongols. Although most of the royal stock was of Mongolian breed, the Song still kept a small number of Fergana "blood-sweating" horses inherited from preceding dynasties. In the wake of the Mongols' takeover, the remnants of the horse stock became diluted into a generous mix of local breeds and eventually became extremely depleted.

The military value of horses was undoubtedly important to the Mongols, who established measures to ensure an ample supply of mounts. Strangely, one of the breeding programmes was to cross Mongolian mares with donkeys to produce mules, which were believed to be superior to horses both as draught and pack animals.

The bulk of the horses was owned by the government and aristocrats. Under Külüg Khan (reigned 1311–1320), the government provided resources to set up shelters for the horses to protect them from bad weather and sourced salt from across the country to keep them healthy with essential minerals in the warmer central, southern and eastern regions.

To maintain their plentiful supply of healthy equine forces, the government occupied or confiscated land from the local populace for horse rearing. Horse farms were set up not only in the north, which was the traditional grazing region, but also in Beijing, Shandong, Yunnan and Henan, which were at high altitude, had cold temperatures, lush grass, fertile soils and abundant water.

Horses were grouped by age and type, from geldings and mares to foals, each of which was raised individually by specialized shepherds.

An annual audit was conducted in September and October. If horses died from disease, the shepherds responsible had to compensate the loss with a big stallion for every three horses that died; a two-year-old yearling if two horses died; and a ram for one dead horse. If the shepherds were unable to compensate the losses with horses, they could provide goats, camels and cattle of the same market value.

The government would also buy horses from the local populace but often at a low price. Kublai Khan (reigned 1260–1294) once put an order out to buy 25,000 horses in 1261, while Gegeen Khan (reigned 1320–1323) commanded the purchase of 100,000 horses in 1320. The policy reached its pinnacle in the late Yuan dynasty. Under Toghon Temür (reigned 1333–1370), if a family had 10 horses, they had to sell two to the government. This reflected the increasing military demand for horses and the failure of official horse farms.

If the government could not purchase enough horses, they would simply confiscate them from the local populace. Mongols, foreigners and officials could keep part of their horse stock, wealthy people would relinquish one-third, but the Han, which included the Han Chinese, Koreans, Jurchens, etc., and Southerners, which included both the Han Chinese and ethnic minorities in the south, had to surrender all their animals. The local populace thus kept fewer horses in order to minimize the loss.

Horse slaughter was forbidden during the Yuan dynasty, even for purposes of ceremonial banquets for senior officials. Sick and old horses had to be certified as being of no use before slaughter. If caught secretly slaughtering horses or cattle, each perpetrator would be sentenced to 100 strokes of the paddle and to a fine, which in turn would be given to the informer. Paddling was the major form of corporal punishment used for horse-related offences and involved beating the back, buttocks or legs with a paddle. Caning was another major form of corporal punishment for less serious crimes, when the buttocks would be beaten with a light bamboo cane. Neighbours who knew the offences were being conducted but did not report them to the authorities would get 27 lashes of the bamboo cane. Officials who failed to pursue offenders would be given 57 lashes. Witnesses who did not report crimes in order to blackmail the perpetrator were punished with 77 strokes of the paddle. If horses or cattle died of disease, the owners had to submit their horns and/or sinews to officials. If the animal's skin and flesh were not retained for personal use and were sold, the owners had to pay a tax, or they would be pursued for tax evasion. Not submitting the horns and/or sinews to officials would earn the miscreant 27 strokes of the bamboo cane for one set, 47 strokes for more than five sets and 67 strokes for over 10 sets. In addition, each offender had to pay a fine equal to the value of the related animal parts, and the fine, once more, was paid to the informer.

For slaughtering official horses or cattle, the crime's ringleader would receive 107 strokes of the paddle and any accomplices would receive 87 strokes. Assisting in the slaughter of official horses or cattle would earn a sentence two levels lower than that handed out to the ringleader.

Harsh punishments were also handed down for horse stealing. Those convicted had to compensate nine heads of livestock for each horse stolen. First-time offenders would get 97 strokes of the paddle and two and a half years of compulsory penal servitude; accomplices would receive 87 strokes of the paddle and two years' compulsory penal servitude. Second-time offenders would get 107 strokes of the paddle and be sent for military service in remote border areas. In remote areas, offenders unable to afford compensation would also have to do military service on remote borders.

In the late Yuan dynasty, the punishments for stealing horses and cattle became much harsher. Robbery of horses would result in capital punishment. Anyone stealing a horse would get his nose slit or cut off. Stealing a donkey or mule would result in being indelibly tattooed on the face or forehead. If the perpetrator reoffended, he would lose his nose. Stealing sheep, goats and pigs meant an indelible tattoo on the nape of the neck;

second-time perpetrators would receive tattoos on their foreheads or faces; third-time perpetrators would lose their noses. After that, a further offence would result in capital punishment.

The Yuan dynasty saw the third high point in China's equine numbers. However, while the Yuan focused on breeding Mongolian horses, which with their excellent stamina rather than speed and appearance had proved themselves in numerous victories across the land mass of Asia and Europe, they decimated or ignored most of the other superior imported breeds, which had been highly valued in preceding dynasties. In the eyes of most Han Chinese, the Mongols – the strongest cavalry force in history – ended up degrading the overall horse quality.

THE HORSES OF KUBLAI KHAN

During Yuan times, extensive trade between East and West resulted in a rich and colourful cultural diversity. Marco Polo (1254–1324) is probably the most famous Westerner who journeyed on the Silk Road to China. The Venetian trader and explorer spent 17 years in China and became a confidant to the Great Khan. In the book *Il Milione* (The Million, or more commonly The Travels of Marco Polo), many of his references are associated with horses.

According to the travelogue, the Great Khan kept a pool of 10,000 snow-white mares that supplied milk to the Mongol court. The court was open every year for an inspection tour of the herd.

The travelogue also revealed how the Khan's formidable and efficient mail system operated. It consisted of roughly 1,500 sumptuous and well-guarded postal relay horse stations set up every 40 to 48 kilometres along all major roads across the Mongol Empire.

Every station held more than 400 horses, all of which had been surrendered by the local populace as part of the taxation system. Half of the horses would be grazing while the other half would stand by. For express services, messengers travelled with a bell hung to their waists so that people would give way and the station ahead would prepare for their arrival. They would recuperate, change to refreshed horses or pass the mail to another messenger.

Under this relay system, an express mail could cover as much as 260 kilometres round the clock.

In 2009, an annual event modelled on the system was introduced in Outer Mongolia for extreme riders. Called the Mongol Derby, it is considered the toughest and longest horse race in the world. It covers 1,000 kilometres of often rugged, isolated and inhospitable terrain (the exact course changing each year), with 25 relay stations and rest stops.

The international riders spend 13 to 14 hours a day in the saddle on continuously replaced and unfamiliar semi-wild horses, which are not always cooperative. Despite previously proven horsemanship, only about half of the riders (men and women) finish. The cut-off time is 10 days. The event supports local Mongolian herders (who hire out their horses and offer yurt accommodation as a camping alternative), helps maintain the Mongolian breed and raises funds for a chosen charity.

MING DYNASTY (1368–1644)

In the early Ming dynasty, the horse's essential role in transportation was downplayed by an increase in the development of shipping. Nevertheless, its military role remained significant; the horse remained crucial for providing military security, although it was always in short supply.

Since the Mongols had impoverished China of most of its quality breeding stock, the Ming had to import more than 10,000 horses every year throughout the course of the dynasty.

From 1405 to 1433, the Admiral-eunuch Zheng He led seven naval expeditions to the South China Sea and the Indian Ocean. Zheng He's fleets consisted of between dozens and hundreds of large treasure ships and smaller ships. Among the smaller ships were specialized ships to carry horses. These ships had eight masts and were 103 metres long and 42 metres wide. As well as horses, they transported tribute goods and timber for repairs.

In his last four voyages from 1413 to 1433, Zheng He reached as far as Hormuz, a bustling Arab trading port in the Persian Gulf, and later East Africa. He bought and received tributes like giraffes, zebras, lions, leopards, camels, ostriches, precious stones and metals, pearls, golden amber, incense, ivory, rhino horns and above all, Arabian horses.

The "tea for horses" trade, which had been very active during the Song dynasty, continued into the Ming. In order to be able to get more horses, the Ming controlled tea production in an attempt to maintain tea prices at an artificially high level. From 1404 to 1423, the Ming's equine population, which had numbered 50,000, soared to about 1.6 million through trade and tribute.

Gunpowder, which had already been used by the Song, was widely employed in warfare alongside infantry and cavalry in the Ming dynasty.

The early Ming saw a hoard of new designs and advancements in gunpowder weaponries, which included exploding cannonballs, land mines, naval mines, multi-stage rockets and hand cannons with up to 10 barrels.

By the mid- to late Ming, the Chinese instead relied on imported artillery and firearms from Europe. To make the most of the imported weapons, they also replicated the technologies, e.g. the breech-loading swivel gun from Portugal and the *hongyipao* from the Portuguese in Macau.

In 1592–1593, the Ming cavalry, backed up by artillery and hand cannons, rebuffed the Japanese invasions of Korea.

Military modernization however reversed in the Qing, which became rather reliant on weapons like the lance, spear and sword, and so forth. Their abandonment of fire power widened the military gap with Western powers.

QING DYNASTY (1644–1911)

In the waning years of the Ming, the alien Manchus became a significant force with which the Chinese found themselves having to contend. While the Ming armed their military forces with advanced equipment, the Manchus used mainly arrows and the arquebus, but better tactics.

In the critical Battle of Sarhu in 1619, Nurhaci, founding father of the Qing dynasty, led 60,000 Eight Banner troopers against the enormous expeditionary Ming force made up of 140,000 soldiers. With their greater mobility, superior fighting power and higher morale, the Manchus overcame the Ming in five days, slaying 310 Ming officials and 45,800 Ming soldiers, pillaging a total of 28,000 horses, donkeys and camels as well as over 20,000 items of gunpowder weaponry. After this defeat, the Ming seriously lost ground and soon faced total collapse.

In 1644, the Manchus, refusing to abandon the territory they had gained, swept into the heartland of China after being invited to help subjugate a rebellion in Beijing. That same year, they overthrew the crumbling Ming dynasty and proclaimed the Qing dynasty.

Like the Mongols, the Manchus had developed a long tradition of equestrian prowess. The founders of the Qing attributed their victory over the Ming to their expertise in mounted archery. Performance in the skill formed part of the public examinations to select military officials.

Mulanquimi (Figs. 1.13 and 1.14) was their most festive horseback hunting exercise and held every autumn. The tradition was revered by the early Qing emperors and carried forward until it was finally abolished in 1824. Kangxi was among the greatest emperors of the dynasty. He had been well trained in childhood in traditional Manchu martial arts and horsemanship. The emperor was quite at home on horseback and very keen on hunting. During his reign, 70,000 horsemen and 3,000 archers accompanied him and senior officials in taking part in Mulanquimi. He reckoned that a trooper could be skilled at archery only if he was also skilled at riding, and that hunting was a good exercise to enhance both of these skills.

In the first half of the Qing dynasty, the cavalry forces were strong enough to help expand the empire and protect it against foreign invasion along China's western and northern borders.

However, in the 19th century, towards the end of the dynasty, foreign encroachment and gunboat diplomacy prevailed and the cavalry of the Manchus was rendered virtually obsolete by the superior weaponry and technology of Western powers.

In a similar way, the demise of the cavalry was made clear in Europe during World War I when the force of the horse in modern warfare was replaced by gunpowder.

A CHANGING CULTURE

In tandem with the rich and colourful past of China's horses, the country's borders changed dramatically over time. The restive minority regions on China's periphery, with their endless expanses of grassland and long-established equestrian traditions, were not always included in the domain. At times, China had no access to these areas and was without suitable forage and habitats; the central plain was anything but a horse-breeding region and pastoralism could barely subsist.

Ancient China was an agricultural-oriented society in which people depended on farming for their livelihoods. Horse rearing needed more space and greater resources. A single horse consumed as much grain as an ordinary family of six, while its economic benefits were less than those of an ox or a sheep. More pertinently, oxen rather than horses served as the indispensable device for the survival and development of mankind. As such, it was difficult for horses to become an intrinsic part of mainstream Chinese culture.

In spite of that, the equestrian heritage of China still radiates amid the history of the world through the civilization's intellectual and artistic creativity. It is a mark of the ingenuity of the Chinese that they developed the stirrup, the breast-strap harnessing system and the horse collar, three of history's most important equestrian inventions. Likewise, they bequeathed a rich range of equestrian-related artistic and cultural traditions to posterity.

CONTEMPORARY DEVELOPMENTS

While the horse may have become diminished in its military, educational and ritual value in modern China, its relationship with the Chinese is still close and the hoof print is still important. In 2019, China's horse population had dropped to 3.7 million from 5.6 million in 2009, but still accounting for about 6 per cent of the world's equine population and encompassing 47 distinct breeds. Following the rise in prosperity in China, this majestic animal has started a new journey.

Fig. 1.13 Giuseppe Castiglione. *Deer Hunting Patrol: Sounding horns to lure deer.* 267.5 x 319 cm. Palace Museum, Beijing. This painting illustrates Qianlong's enormous hunting cavalcade entering the Mulan hunting ground in northern China in 1741 when the emperor first attended Mulanquimi. The third rider at the head of the cavalcade on a white horse and with a red quiver is the 30-year-old Qianlong. In the middle right, members of the emperor's hunting entourage carry horns. Before the hunt, they would imitate the sound of the deer by blowing the instruments to lure them out of the forest. The Mulan hunting ground was located about 450 kilometres north-east of Beijing and was a natural habitat for many species of animals, especially deer.

Fig. 1.14 Mulan hunting ground as it is now.

China's booming economy in recent years has produced many nouveau riche and bourgeoisie who are embracing the opulent lifestyles and lavish buying habits of some of their Western counterparts.

Dining in smart restaurants, buying hi-end goods like designer handbags, jewellery, perfumes and watches are common ways for them to spend their newfound wealth. China had already surpassed Japan in 2012 as the largest market in the world for luxury goods.

LEISURE AND BUSINESS

As Western consumerism spreads, having a classy hobby is another way of flaunting success. Emblematic of status comparable to owning a Mercedes Benz in the soaring economy, equestrian sports have become a fashion statement among the burgeoning elite and expatriates in China. According to Chang Wei, the former General Secretary of the Chinese Equestrian Association and current Chairman of the Development Planning Committee of the Asian Equestrian Federation, in 2015, China had some 600,000 equestrian sports enthusiasts who practised twice a week and more. Every year, more than 200 international and local tournaments take place in China. The Olympic Games of 2008, the Asian Games in 2010 and the FEI World Cup held in 2015 are among the major events staged in recent years.

The hobby has also spawned a host of business opportunities, including riding schools, horse farms, equestrian magazines and riding gear, all of which are expanding apace and growing into promising markets. A new wave of riding clubs, polo clubs and breeding farms has proliferated in an unprecedented and unexpected manner. In 2015, there were over 800 clubs across the country. These are located in Beijing, Tianjin, Tangshan, the north-east, the Yangtze River Delta, the Pearl River Delta, Shanxi and Sichuan, to name just a few. Beijing, Shanghai and Guangdong have the greatest numbers.

While most clubs may look a bit shabby, some have reached an international standard boasting of imported breeds, flashy stables, paddocks, reception areas and other facilities in their promotional materials. This makes China an attractive target for European stud farms seeking opportunities in the nascent market. For example, Irish Thoroughbreds and Russian Kapakins have now been introduced in Inner Mongolia to interbreed with Mongolian horses. In 2008, only 573 horses were imported into China. In 2019, the number increased to 1,849; the Netherlands topped the list for these with 861 heads at an average customs value of US$9,142 each, and Outer Mongolia followed with 280 heads at US$1,300 each.

With its widespread and sophisticated use of horses in daily life and for leisure activities over many centuries, the Western world possesses a more advanced and extensive knowledge in training and breeding, equitation, diets and veterinary care. In recent years, while Chinese guests have already been on the invitation lists of sport horse auctions in Germany, France and Australia, auctions of imported horses are also staged in China. Inspection tours by interested Chinese to leading overseas equestrian countries are becoming more frequent and are aimed at understanding each other's equestrian standards and looking for opportunities for cooperation. More and more seasoned riders and trainers come to China, sharing their experience and providing training for enthusiasts.

The import of premium horses, quality riding equipment and equestrian consultation services are beginning to take place in modern China, helping to balance the enormous trade surplus complained about by the West at times.

China is well known as a global factory, so there is little wonder that much of the world's equestrian gear is produced there. The industrial zone on the outskirts of Shanghai is home to a vast pool of plants making every type of equestrian product from brushes and breeches to boots and whips, from premium international brands to mainstream local brands. Equestrian trade fairs are held every year with a wide range of products and services available to view and compare, including horse trading, equestrian apparel, forage and veterinary supplies, horse appliances and competition equipment, horseback sightseeing tourism, equestrian leisure facilities and related products. The events attract thousands of local and international participants from Britain, the United States, Japan, Germany, Spain, Switzerland, Ireland, Egypt and Russia. They are suppliers, buyers, specifiers, users and industry experts, among which a bunch of leading names are included.

Despite all this activity, *Equestrio*, a luxury equestrian bilingual magazine launched in 2008, was finally wound up in 2015. The market essentially does not have enough sophisticated, wealthy or status-seeking horse lovers to sustain the title.

Many riding clubs have also been liquidated as they could not make ends meet. Some clubs sought to control operating costs and exhausted their horses by prolonging their working hours, especially during the weekends, to satisfy an influx of customers. Horse welfare is just not at the forefront of most operators' minds.

Many other issues are also looming. Currently, most instructors do not have formal training and qualifications like those of the British Horse Society; some are just ethnic minority riders. Horse transport lags behind also, unable to support tournaments well. Veterinary standards are not up to what are internationally recognized; the nation does not have full and advanced facilities, enough qualified veterinarians, sufficient and proper Western medicine and supplementary education programmes. Equine insurance cannot be developed because of an absence of professional evaluation.

REAL HORSEMANSHIP

Equestrian sports in China may look posh, more like a status symbol than a pure physical activity, nevertheless, real horsemanship does exist, and outstanding horse riders and horses can still be found.

Made up as they are of vast tracts of desert, mountains, steppes and plateaus, the northern, western and southern border areas of China are too high, too cold and too dry to support a dense agricultural population. This region, which includes Inner Mongolia, Qinghai, Tibet and Xinjiang, is close to the Eurasian Steppe, which is the homeland of horses and has a climate and terrain especially suitable for the animal. For centuries, a complex mosaic of ethnic minorities settled in the areas developing many and varied pastoral-nomadic traditions. Nowadays, modernization flares across the region and modern modes of transportation are being increasingly adopted (Fig. 1.15). Satellite dishes and wind generators in the Inner Mongolian steppe, and modern consumer electronics like televisions, cookers and mobile phones owned by Xinjiang families, are by no means uncommon.

However, these modern elements cannot break the inextricable link between horses and the nomadic tribes, whose equestrian heritage and passion for their animals remain unabated (Figs. 1.16 and 1.17). Although no longer a necessity for survival, horsemanship is always venerated as a matter of cultural pride.

The Naadam Festival has been the Mongolians' biggest fair for over 700 years. The word "Naadam" in the Mongolian language means "entertainment" or "game". The harvest fair is celebrated on the fourth day of the sixth lunar month (between July and August in the solar calendar) when drinking water is plentiful, the pasture is lush, and the livestock have become strong and stout. Showcasing the "Three Manly Virtues" of the Mongol men: horse racing, archery and wrestling, the festival is open to participation by people of any ethnic origin and religious belief. A notable feature is that herdsmen arrive wearing their traditional costumes from different corners of the country. Nowadays, in addition to cross-country horse racing, more equestrian activities are held throughout the year, such as lassoing (Fig. 1.18), horse-drawn sleigh racing (Fig. 1.19) and ball tournaments. Riders compete along with horses for speed, stamina, strength and bond of trust, demonstrating virility and equestrianism.

Likewise, the Tibetans have the Kyagqen Horse Racing Art Festival of Nagqu and the Darma Festival of Gyangze.

Apart from celebrating their traditional equestrian festivals, herdsmen also enter modern equestrian competitions. At times, they can be seen edging out rivals from other ethnic groups and winning rosettes in different disciplines.

Fig. 1.15 Herding livestock by motorbike is prevalent in Inner Mongolia, especially among young shepherds. Sheep have become used to the motorcycle horn as a herding call.

HORSE RIGHTS AND WELFARE IN CHINA

EARLY RACING EXPERIENCES

The Huajun Horse Racing Club (also known as the Tongshun Race) was once the largest racecourse in Beijing with reportedly 2,000 Thoroughbreds imported from Australia kept at its adjacent stud facilities. The racecourse opened in 2002 as an investment by a Hong Kong businessman in the expectation that the largest untapped betting market in the world was set to legalize gambling. Since gambling was forbidden, the races were planned to be operated on a sports-only basis. However, the business was shut down by the Chinese government in October 2005 because of alleged on-site gambling despite a "Strict Forbidden Order on Horse Racing Gambling" released by the central government.

This investment backfire was paid for by the innocent animals with their lives: the club reportedly put down 500 horses. It later claimed that the number was between 120 and 150, and was in line with international practices to cull weak, injured, disease-ridden and barren horses. Still keeping alive the hope of horse racing being allowed at some point on the horizon, the cash-strapped racing club continued to keep some 2,000 horses. Worse was to follow.

During August and September 2009, a furious land dispute broke out between the club and local villagers, both of whom had entered into a 30-year lease in 1997. The dealing that went on behind this is somewhat murky, but it seems the lease had not been negotiated by the villagers, who later found the rent they would obtain was too low to make a living. A dispute also arose over some areas where land use rights were unclear. The villagers

asked to be able to raise the rent for the land, but the club refused to bargain. Although the police were called in, they couldn't interfere because no violence had taken place. Having failed to reach an agreement with the club, the villagers turned out in considerable numbers to block its entrance, restricting access by staff and feed deliveries to the 160-hectare facility. As a result, more than 30 horses died of starvation and a great number of Thoroughbreds were severely malnourished. Nevertheless, the expensive horses in the club were still fed as usual.

It was apparent that the horses were treated simply as commodities and only the "valuable commodities" were protected. What is more disturbing is that after the ordeal, the club kept 1,750 horses and continued to import mares for breeding!

We all know that the horse racing world is harsh – big business rather than a pursuit – but as far as the main participants in this sorry story were concerned, the horses were just "goods" and ended up being culled when they no longer had a commercial future.

In this deplorable situation, if the owners were unable to find any alternative use for the horses, the only option may have been to euthanize the animals humanely. However, there is no excuse to starve a horse, even for a day! The owners should have mobilized all their resources to have looked after them ALL, not just the expensive ones. In places like Hong Kong, Europe and Australia, which have strong animal welfare laws, starvation could not have happened.

Lack of Animal Welfare Laws

Currently, China does not have comprehensive animal welfare laws. Other than for endangered species, e.g. Przewalski's horses, Tengchong (Yunnan) pony and Jinjiang horses, which are protected, there is no penalty for abusing or slaying animals.

As has happened for a very long time, local governments engage in the mass culling of stray dogs to combat rabies, which is invariably fatal, applying methods which include burying the animals alive or beating them to death. The law formulated to prevent animal epidemics is directed solely at the maintenance of the health and welfare of humans, but not at the welfare of the animals.

Growth of Animal Rights Awareness

Despite the fact that there is little legal ground to act upon, an increasing number of animal lovers across the country unhesitatingly fight against animal cruelty.

In April 2011, 500 dogs were seen being crammed into a truck on the outskirts of Beijing en route to a slaughterhouse. More than 200 animal rights campaigners, mainly from the wealthy, pet-loving and urban middle class, confronted the driver and finally paid RMB110,000 (US$16,000) to rescue the whole truckload of dogs, which would otherwise have ended up being served at a dinner table. Half a year later, a similar case occurred in Sichuan where 400 stray dogs were rescued. In September 2011, China officially banned an ancient dog-eating festival that dates back more than 600 years in Qianxi, a small township in Zhejiang, after an online furore was sparked among animal rights activists. Standoffs between animal lovers and animal exploiters, or the rich and the poor, or the rural and the urbanites, happen over and over again and have generated widespread coverage and discussion in recent years.

Against the backdrop of China's economic boom, swelling numbers of middle class

Fig. 1.16 (Opposite) Mongolian guests attend the opening ceremony of the Ordos Museum in Inner Mongolia in 2011 with their horse.

Fig. 1.17 (Overleaf) Mongolians walk with their horse around the interior of the Ordos Museum. The horse wears two pairs of blue boots in keeping with the colour of its owners' clothes. Bright blue is a common festive colour in Mongolian culture.

Fig. 1.18 Lassoing in a winter Naadam Festival.

city-dwellers are emerging with both money and time for animals. Education and the influence of Western culture have also helped change their concept of animals such that they no longer consider them just commodities, but see them as friends and human companions with personalities and cognition. From time to time, they initiate online petitions against animal cruelty, post videos

DRAFTING OF GOVERNMENT LEGISLATION

In September 2009, China unveiled its first ever draft animal protection bill governing acts of cruelty to animals. It covers legal protection for different types of animals – wildlife, pet, farm, entertainment and laboratory animals, and regulates animal abuse prevention, healthcare, transportation, slaughter and the eating of cats and dogs.

Because of tradition, poor educational levels, culture, values and economic background, some people, especially those from the remote rural areas, cannot accept the concept of animal protection or animal welfare. In their minds, human interest is the overriding priority. The gap between the rich and poor in China is still very wide. Peasants in the rural areas are struggling with poverty and look for more protection and welfare for themselves. Therefore, the draft bill puts greater emphasis on anti-cruelty than anything else, as this is considered easier for the public in general to accept.

However, the draft has yet to be finalized and little news of it is heard.

PROMOTING HUMANE HUSBANDRY

Although the draft animal protection bill is yet to be passed, promotion of humane slaughter began over a decade ago in 2007.

As well as China's more sophisticated middle class, the animal products industry in the country is also a big espouser of better animal welfare. There are several reasons that it is in the industry's interests to be so.

Improvement in animal welfare is actually indispensable for the industry if it is to achieve sustainable development, most pertinently in helping remove trade barriers faced by China in exporting animal products like meat, wool and feathers.

The poor animal welfare standards upheld in the meat industry have long been complained about, especially the cruelty found in animal husbandry, the methods of killing animals and the delivery of live animals. Since more than 100 countries have a legal framework on animal welfare and the World Trade Organization also has rules covering the issue, China is prevented from exporting meat products, except for rabbit and cooked poultry, to

and photographs showing the maltreatment of farmed animals, criticize inhumane practices, unite to rescue animals and in doing so move animal welfare in China forward.

many overseas markets, especially those of the European Union, which applies a higher animal welfare standard.

Pork meat production in 2019 amounted to 42 million tons, while horse meat production reached 200,452 tons in 2018. Both represent the highest figures in the world. However, the total amount of pork and horse meat exported in 2019 was 26,630 tons and 129 tons, respectively. Japan was the only destination for the horse meat. Exports to the international meat market from China are therefore minimal.

In February 2007, a humane slaughter programme was jointly developed by World Animal Protection, formerly known as the World Society for the Protection of Animals, and Beijing Chaoyang Anhua Animal Product Safety Research Institute.

Fig. 1.19 Horse-drawn sleigh racing in a winter Naadam Festival.

The objectives of the programme were to promote rules and standards for humane slaughter and to provide training to industry participants for the implementation of humane slaughter. The programme planned to provide training to 20,000 abattoirs across the country. As of the end of 2015, more than 6,000 people from nearly 1,200 companies and 15 universities had received the training. The non-profit animal welfare organization also works with major suppliers in provinces nationwide to advocate sustainable, humane and economically viable farming in China. Encouragingly, the participating farms have found themselves making more money as the health conscious and compassionate middle classes are willing to pay extra for produce that is both safe and ethically sound.

PIGS LEAD THE WAY

The humane slaughter equipment industry has also received a boost, growing alongside the promotion of animal welfare.

In August 2008, the Ministry of Commerce formulated regulations on the humane handling and humane slaughter of pigs, and a set of standards was accepted into the guidelines on the technical criteria for the humane slaughter of the animals. The new standards suggest playing music to the

Since February 2010, the city has enforced regulations for the humane slaughter of pigs. These prohibit pigs from being moved, fed or forced or induced to drink a great amount of water in the 12 hours before slaughter. They have to be left alone to rest and relax free in the pen before being killed. Three hours before slaughter, water must be withheld. Workers must keep a close eye on the animals to make sure that they are safe and in good shape. If it is found that they have been given less than 12 hours of rest, the enterprise will be subject to a penalty of between RMB20,000 (US$2,930) and RMB50,000 (US$7,300). If forced or induced drinking occurs, all the pork and pork products will be destroyed and the enterprise will have its operations suspended or even its slaughter licence revoked.

According to people in the meat industry, resting and starving a day before the slaughter, after long-distance transportation, can help eliminate fatigue, enhance drainage and detoxification and avoid contagious diseases spreading among pigs. Even though this routine increases cost, a better taste to the pork will be ensured and water-logged pork will be prevented from entering the market. Both factors will help improve the image of the industry.

In October 2015, the Guangzhou authorities proposed a bill on pig slaughter as well as pork and offal transportation and logistics, in which humane slaughter is encouraged.

In August 2016, Shandong, which supplies more than 20 per cent of domestic chickens, issued guidelines, albeit not mandatory, for the humane slaughter of poultry, with an aim to minimize the birds' pain, fear and stress before and during the slaughter. According to the guidelines, the transportation time for chickens going to slaughter should be limited to a maximum of three hours; temperature and hygiene should be observed; and dragging chickens by one of their wings or legs should be prohibited. Upon arrival at the slaughterhouse, the birds should be given a break to relax. On the conveyor carrying them to be anaesthetized by gas inhalation or other methods, they should be provided with massaging pads on their breast to help calm them down. Chickens are ready for slaughter only when they fall unconscious.

pigs and patting them to help them relax before having their throats slit. It also includes shortening the interval between stunning and bleeding the pigs from 30 seconds to 15 seconds, to reduce the pain and fear experienced by the animals. The practice also helps to improve the quality of the pork. If bleeding is delayed, the animals' blood pressure will increase and cause rupturing of blood vessels and muscle haemorrhage. This extra blood in the tissue hastens decomposition and is more likely to lead to wastage.

Zhengzhou in Henan was the first city to implement humane slaughter in 2007. Henan is the second largest pork-production province in China, supplying 3.4 million tons of pork in 2019.

Effort has also been made recently to regulate the veterinary profession and set training and qualification standards. More positively, on February 24th 2020, following the outbreak of coronavirus in Wuhan in December 2019, which resulted in the global pandemic, the Chinese authorities imposed an immediate ban on eating wild animals, including all the related hunting, trading and transportation. Later, on April 8th 2020, the Chinese government issued a draft policy which would likely result in ending the consumption of dog meat, affirming that the animal has evolved from traditional livestock to a human companion. Shenzhen, a major city in southern China, then took the initiative to prohibit officially the eating of dogs and cats, effective May 1st 2020. While there is still no general legislation on animal protection, these latest events all represent great strides in the advance of animal welfare in China, and it is hoped they can be extended soon to encompass horses and other farm animals.

Voice of the Equestrian Class

Equestrian pursuits are a magnet for the ranks of the middle and upper classes whose economic, social and political clout is capable of exerting pressure on the government. To help ensure the protection and welfare of horses as a whole, this group of people should act to demonstrate their strength, their civilized attitudes and their true love of the world's fastest distance-running quadrupeds. Knowing that by doing so they show respect for the welfare of all horses should increase their enjoyment of their own equestrian pursuits and ultimately benefit all other domesticated animals.

GOING FORWARD

The Chinese Equestrian Association has been a member of the Federation Equestrian International (FEI) since 1982. In 2007, the FEI brought its jumping and dressage examination system to China for the first time. More and more events follow the FEI rules. The 2011 China League of the FEI World Cup jumping competition was the first FEI jumping competition held in China. Taking place in Beijing, it featured three Concours de Saut International (CSI) events. The CSI is a ranking system for equestrian show jumping competitions approved by the FEI.

It is believed that one day, when the conditions in China become more favourable with more qualified horses, quality veterinary care, sophisticated breeding and training programmes, and, most of all, horse protection and welfare laws in place, Chinese riders will flex their muscles in international arenas, competing with the world's top contestants.

Fig. 1.20 The many different ways the word horse can be written in Chinese.

CHAPTER TWO

THE HORSE IN ART AND EQUESTRIAN ARTEFACTS

INTRODUCTION

While the Earth's fastest distance-running quadrupeds are much revered in China, they also hold great prominence in the artistic realm as we see them traversing time and space over the millennia of Chinese cultural development.

Since soon after the horse was first ridden on the mainland 5,000 years ago, the Chinese discovered the many practical uses to which the animal could be put, from warfare and transportation to education and the means of making a living. Impressed by its aesthetic beauty, they explored the animal further, and in doing so the horse became a timeless and much-loved subject of artistic creation. Equestrian artefacts in bronze wares, terracotta and pottery; paintings, murals and scrolls; and stone carvings, all with distinct national characteristics, were created to serve different purposes in different dynastic periods. Inspired craftsmen added lustre to the Chinese people's artistic achievements with each succeeding epoch.

From the Warring States period (475–221 B.C.) to the Tang dynasty (618–907), most artefacts were commissioned by gentry from the ruling classes as a form of dedication to publicize and eulogize loyalists and martyrs, and to stigmatize traitors and invaders. Horses in warfare were the major subject for such commissions, epitomizing valour, energy and power. While the brutality of these dynastic battles has long since been lost to time, the glory of the horse remains immortal, gracing the tombs of such great heroes as General Huo Qubing and Emperor Tang Taizong.

Native Chinese horses, noticeably more diminutive and stocky, are commonly found in equestrian works of art until the early Han dynasty. From then on, more and more imported horses, especially the blood-sweating horses, had arrived in China. Blood-sweating horses were so magnificent that emperors, aristocrats and artisans were completely mesmerized by them and made them the subjects of equestrian works of art, replacing native breeds. As such, horses in Chinese artefacts were then portrayed as having slim and long limbs, graceful heads and bodies, elongated necks, silky coats and fine skin.

During the period of the High Tang dynasty to the Song dynasty (960–1279), the battle-related theme in art was replaced by the pure pleasure of equestrian pursuits, a stylistic change which reflects the prosperity of the time. Tri-coloured glazed pottery horses along with such as the "Silver Saddle-flask with Gilded Dancing Horses" portray the vibrant scenes of the most dazzling historic period in Chinese history. Its *Five Tribute Horses* travel a long way to the Song court, while *Night Shining White* laments that there is little enjoyment in being a royal steed.

In the meantime, the blood-sweating horse was still the leading figure in equestrian works of art but its slender frame had ballooned after the High Tang dynasty when a voluptuous image dominated the aesthetic standard across the different creative mediums. Even after the Tang, the dynasty's grandeur and flair continued to be carried on the back of patriotic painters yearning for this former golden age in Chinese history. The Song dynasty, with its weak military force, was consistently vulnerable to the encroachment of contiguous states, and nostalgia for the Tang era thus permeated society all the way up to the Song court. As a result, replication of Tang artefacts was not uncommon and even encouraged by the ruling class. Also thanks to this encouragement for replication championed by Emperor Huizong (reigned 1102–1125), we have a chance to see those Tang artefacts which had already vanished.

In the third stage of artistic development, from the Yuan dynasty (1279–1368) to Modern China (1911–present), craftsmen sought to express themselves through their artefacts. As the famous saying of the Tang essayist Han Yu (762–824) tells us: "There are always excellent steeds, but not always a Bole, the legendary connoisseur of horses." Many scholar-painters used brushstrokes and horses to convey their grievances over having their talents neglected. Gong Kai's *Emaciated Horse* embodies a patriot who serves an alien court, while the horses in *Fat and Lean Horses* take to the streets, criticizing social injustice. The animals in *One Hundred Horses* are more than just fixtures of the imperial garden: they open a curtain onto the hybrid style of East and West. Finally, *Galloping Horse* expresses agitation over the fate of the nation.

In the 16th and 17th centuries, Jesuits rolled out their missionary work in China while introducing Western science, mathematics, astronomy and visual arts to the Chinese imperial court. Western arts at that time pursued a realistic depiction of the material world, applying graphical perspective to give a more obvious three-dimensional depth to artefacts. With their influence, the appearance of horses in Chinese art also became more realistic; no longer did the art reflect a symbolic and distorted graphical image.

No matter what messages the art delivered and what images of China's horses were displayed, the passion of the Chinese and their strong bond with this fascinating and inspiring creature run very deep, as proven by the country's rich and amazing legacy of artistic representation.

QIN DYNASTY (221–206 B.C.)

TERRACOTTA HORSES
EMPEROR QIN SHI HUANG'S MAUSOLEUM SITE MUSEUM, XI'AN, SHAANXI

The subterranean terracotta army, which lies near the necropolis of Qin Shi Huang, the first unifier of China, in Xi'an, Shaanxi, is among the most spectacular archaeological finds of the 20th century. Created 2,200 years ago as imperial guards to attend the emperor in his afterlife, these thousands of figures appear in the splendour and verve of the period in which they served.

The discovery of the terracotta army was made in 1974. Since then, 7,000 figures, tens of thousands of bronze weapons and other objects, and most of all, 516 chariot horses and 116 cavalry horses have been unearthed. Before the Qin dynasty, pottery figures created during the Warring States period were small and roughly made and were fired at low temperatures. The terracotta horses in Qin Shi Huang's army are all life-sized and are mainly cavalry and chariot horses.

Just like the terracotta warriors, the horses were produced from moulds. They all stand at 172 cm from toe to head and 133 cm from toe to shoulder. The cavalry and chariot horses measure 203 cm and 210 cm in length, respectively. The horses are very close in size to that of war horse skeletons excavated in a nearby stable pit. The *Qin*

Code prescribed that the war horses should stand at least 133 cm tall, and be well trained and obedient; otherwise, the officials in charge of them would be punished. The terracotta horses attest to the strict selection criteria.

Nevertheless, each horse looks a little different, varying in facial expression or the features of the ears, eyes, forelock, nostrils, muzzle and mouth, which were hand-finished.

The cavalry is located in Pit 2 where some 116 armoured cavalrymen stand in front of their horses and hold their reins (Fig. 2.1).

Each cavalry horse is engraved in detail with a saddle, a girth, a crupper and a saddle pad. The saddle is more advanced than those used in the Warring States period, with a pommel and a cantle, albeit low. The tack is also very similar to the equestrian finds in the Pazyryk burial site in Siberia. Meanwhile, no stirrups were carved on these horses as they had not yet been invented. Once in use, stirrups could help troopers mount and they enhanced balance in the saddle, thus freeing the riders' hands to fight. Without stirrups, the ability to fight from the saddle was limited.

No short-range weapons used in hand-to-hand combat, like the spear or sabre, were found with the cavalry in Pit 2. Instead, excavations produced crossbows and arrows, which were used to fight from a distance. Since the cavalry did not have a very secure seat in the saddle, it could not fight the enemy at close range. It is thus deduced that cavalry at that time took a less crucial role on the battlefield, but monitored the enemy and attacked their food transportation units and other supplies.

The chariot horses are not depicted with any engraved gear and, unlike the cavalry horses, their tails are braided (Fig. 2.2).

Fig. 2.1 Terracotta cavalry horse.

Fig. 2.2 Terracotta chariot horses.

These quadruped figures also reveal that equine castration was performed over 2,000 years ago. Yuan Jing, an archaeologist with the Chinese Academy of Social Sciences, pointed out that the chariot horses have penises but no testicles, while some cavalry horses, like the chariot horses, also have no testicles, but that some are intact. A gelding is more manageable and easier to train, making it safer, gentler and potentially more suitable as a war horse. Geldings account for the majority of the terracotta horses, indicating that the quality of the horse rather than its quantity was key in the Qin's equine policy.

The terracotta horses are dumpy and stocky, with large heads and robust hooves. Their appearance, especially in size, resembles that of the Hequ horse, which was the major breed of China's western hinterland at that period. The figures were fired at approximately 1,000°C. To prevent any cracking of the clay during the firing process, the horses were made with a hole in their flanks. After firing, the figures, horses and warriors, were painted with bright pigments, mainly vermilion, claret, pink, green, purple, blue, azure, yellow, orange, white and ochre, and covered with lacquer. However, these colours have since been lost to desiccation and oxidation.

The terracotta horses were exquisitely made. Their images were vividly created, each overflowing with raw energy and rustic charm. The big round eyes manifest a mighty spirit; the raised heads and tails articulate alertness; the widening muzzles demonstrate vigour; while the sturdy bodies symbolize fortitude.

HAN DYNASTY (206 B.C.–A.D. 220)

HORSE STEPPING ON A XIONGNU SOLDIER
1.68 x 1.9 M

MAOLING MUSEUM, XINGPING, SHAANXI

(FIG. 2.3)

Emperor Wudi's reign represents the zenith of the Western Han dynasty (206 B.C.–A.D. 8). The Xiongnu were a nomadic people from North Asia

Fig. 2.3 Horse Stepping on a Xiongnu Soldier.

who had been constantly making inroads into Chinese territory during the Han dynasty. As this encroachment became more pervasive, General Huo Qubing, then only 18 years of age, was sent to fight against the Xiongnu. He was a military genius and six times defeated the enemy in the Qilian Mountain area. Huo died at the very young age of 24 and to commemorate his brilliance as a general, a large tomb in the shape of Qilian Mountain was established on the peak.

The human and animal stone carvings which decorate the tomb in Shaanxi are the earliest and best-preserved stone carvings to be found in China. They are of great artistic value in the history of sculpture. The monumental and evocative *Horse Stepping on a Xiongnu Soldier* is widely recognized as the most notable exhibit. The granite sculpture epitomizes General Huo's defeat of the Xiongnu and the mightiness of the Western Han dynasty. The horse, which is the embodiment of a patriotic hero, appears placid, firm and imposing in its trampling of the Xiongnu soldier. The soldier, in contrast, seems urgently animated, a bow in his left hand and an arrow in his right, his hair untidy, and his face showing panic and agony as he makes his last desperate struggle against death.

Granite is a hard medium to work with, but one which can break easily if the pose and gestures are not properly supported. At the same time, carving granite is a subtractive process and cannot be corrected if mistakes are made. As such, the sculpture is succinctly created; the lines and detail are reduced to the minimum. The fluidity of line goes through the horse's legs, hip and neck, resulting in the almost tangible movement and intense impetus.

The other two equestrian sculptures exhibited in front of the tomb are *Recumbent Horse* and *Prancing Horse*.

RECUMBENT HORSE
1.14 x 2.6 M
MAOLING MUSEUM, XINGPING, SHAANXI
(FIG. 2.4)

Recumbent Horse, with its well-trimmed mane and short and sharp ears, appears laid-back and carefree. However, this veteran war horse does not forget its purpose in life. Lifting its head and staying alert, the horse is looking to the front, its ears already pricked. The tip of one hoof just touches the ground while the other is raised slightly, its hind legs already engaged: the horse is all set to carry a valiant cavalry officer into the battlefield.

Fig. 2.4 Recumbent Horse.

The sculpture grasps the exact moment of a recumbent horse when it initiates movement, weaving between its static and dynamic strengths.

PRANCING HORSE
1.5 x 2.4 M
MAOLING MUSEUM, XINGPING, SHAANXI
(FIG. 2.5)

while the hind hooves just touch the ground. The head is raised and a pair of big round open eyes speak of passion. All suggest that this horse, too, is ready to gallop into the battlefield.

Ancient Chinese artefacts do not expressively render the ravages of war. What we can see is only a symbolic depiction rather than realistic bloodshed or the carnage of a battle scene. This is because

Fig. 2.5 Prancing Horse.

The craftsmanship portrayed in *Prancing Horse* follows the stone's natural contours, refined only with a few carefully chiselled lines. The spare but accurate carvings accentuate the uncontrived and effortless beauty of the stone and complement the artistic excellence of the sculpture. In the work, we can see that the forelegs are poised for stretching

artefacts in ancient China were also used as a tool of edification. Confucius championed overwhelming enemies by virtue rather than by violence. A realistic depiction of a fierce fighting display would negate the edifying purpose of the artefact. Therefore, war horses are presented with a vivacious, stately and lofty image, but without any trace of brutality.

BRONZE CANTERING HORSE

34.5 x 45 CM

GANSU PROVINCIAL MUSEUM, LANZHOU

(FIG. 2.6)

This masterpiece was excavated in 1969 in the Leitai Tomb of the Eastern Han dynasty (25–220) in Wuwei, Gansu. Wuwei became famous for this fantastic find, which is among the foremost sculptures in Chinese art and has been used as a tourism icon in China.

The horse's posture is unique and the sculpture immediately identifiable. The statuette is meticulously balanced with its centre of gravity accurately derived to ensure stable support to the whole. Created about 2,000 years ago, the stunning virtuosity of this bronze piece is an extraordinary legacy of China's early artistic achievement.

The enigmatic horse is the ultimate combination of realism and idealism. While on the one hand, the artisan is able to show an astonishing comprehension of anatomy, crafting a startlingly vivid and well-proportioned horse in active movement, he also sets up an imaginary and romanticized scene in which the horse, while engaged in an extended canter, poised yet swift, treads on a bird. Although unable to

Fig. 2.6 Bronze Cantering Horse.

get out of the way quickly enough and trodden to the ground, the bird is not killed and even looks back up to the horse. The dynamic between the agile and gallant horse and the startled bird is compelling. The romantic image of the bird is emphasized against the mighty horse, whose valour and vitality are elevated, enriching the imaginative experience of the viewer.

From its well-built skeleton and musculature, the horse is believed to be a blood-sweating horse from the western regions. The bird, which is always referred to as a swallow, does not have an obvious fork in its tail. As the artisan crafted the horse in extraordinary detail, down even to individual muscle masses and veins, it would be unlikely that he would make a poorly executed swallow. In ancient Chinese literature, swallows were always associated with horses. This may be the reason for the bird being seen as a swallow.

TANG DYNASTY (618–907)

Six Steeds of Zhaoling Mausoleum
STONE RELIEF PLAQUES, EACH 1.7 x 2 M
XI'AN BEILIN MUSEUM, XI'AN, SHAANXI
PENN MUSEUM, PENNSYLVANIA
(FIGS. 2.7–2.12)

The Tang dynasty was an era of unprecedented power and prosperity for ancient China and its most prominent ruler was Emperor Taizong Li Shimin. Honoured by various Turkic nomads as Tian Kehan (the title the nomads gave to Taizong), which translates as Heavenly Khagan, he led the Tang to become the dominant power in East and Central Asia.

Taizong had six favourite steeds which carried him through his military triumphs. To commemorate the emperor's lifetime battle achievements in the unification of China and acknowledge his beloved steeds, six relief plaques were carved from stone and mounted in rows on both sides of the veranda of the northern gateway of Zhaoling Mausoleum where Taizong and his empress were buried.

Without the presence of any peripheral scenery, each relief plaque depicts the individual physical nature and personality of every one of the six horses, articulating its facial expression and riding tack and most of all, rendering the scene of the battle that Taizong and the animal went through.

Yan Liben (601–673), a famed painter of the early Tang dynasty, was commissioned first to portray the six steeds before the sculptors got to work. Taizong wrote a eulogy for each horse, which was transcribed by the great calligrapher Ouyang Xun and carved on a stone slab located at an upper corner of each plaque. The carved eulogies however have been lost with the passage of time.

While the horse sculptures in the Qin and Han dynasties are considered rough and sturdy, the six horse sculptures of the Tang dynasty are seen as being refined and elegant. The six steeds are in low relief, incised with precise and compact lines, producing a shimmering impression of the valour and grace of the animals.

The stone plaques were finally given the name *Six Steeds of Zhaoling Mausoleum*. Guarding the imperial tomb some 80 kilometres from Xi'an for over 1,200 years, they continue to win high praise from both domestic and foreign archeologists and artists.

These six horses all came from the western regions, either from Persia or the Göktürks, and each had a foreign name: Te Le Biao, Qing Zhui, Shi Fa Chi, Sa Lu Zi, Quan Mao Gua and Bai Ti Wu.

Four of the sculptures are housed in the Xi'an Beilin Museum in China, but the other two, Sa Lu Zi and Quan Mao Gua, were looted early in the 20th century and are now preserved in the museum of the University of Pennsylvania in Philadelphia, USA.

Te Le Biao (Fig. 2.7) was the first plaque on the east veranda. "Te Le" was an official title in the Göktürks. "Biao" in Chinese refers to a palomino horse sprinkled with white spots. In the campaign to suppress Song Jingang in 619, Te Le Biao remained tacked up for three whole days, carrying Emperor Taizong very long distances each day. The emperor himself remained armoured during their advance, going without food and water for two of the three days.

The second plaque is of Qing Zhui (Fig. 2.8). "Zhui" in Chinese means dapple grey horse. "Qing" means lime-green but its pronunciation is close to

Fig. 2.7 "Te Le Biao".

Fig. 2.10 "Sa Lu Zi".

Fig. 2.8 "Qing Zhui".

Fig. 2.11 "Quan Mao Gua".

Fig. 2.9 "Shi Fa Chi".

Fig. 2.12 "Bai Ti Wu".

Figs. 2.7–2.12 Six Steeds of Zhaoling Mausoleum (Stone relief plaques).

"Qin" which meant Roman in ancient China. The horse was probably therefore Roman. Qing Zhui is depicted galloping, despite being struck by five arrows – one in the front and four at the back – during the battle with Dou Jiande.

"Shi Fa" is a phonetic translation of the word for horse in the Persian language, while "Chi" in Chinese means bay colour. It was therefore likely to have been a Persian bay horse. However, the pronunciation of "Shi Fa" is close to an official title in the Göktürks, and it is also possible that it was a Turkish breed. The horse carried Taizong on a follow-up campaign to annihilate the residual forces of Wang Shicong and Dou Jiande. Galloping in the battlefield, Shi Fa Chi (Fig. 2.9) is seen with five arrows sticking out from its hindquarters.

The first plaque on the west veranda was Sa Lu Zi (Fig. 2.10). "Zi" means purple. In an ancient Chinese literary text, Purple Swallow is the name of a fine horse. Taizong used Purple Swallow to describe this horse in the form of a blessing. The phonetic translation of "Sa Lu" in Turkic is plucky leader, which was also a senior official title used in the Göktürks. The plaque of Sa Lu Zi is the only one with a human figure.

In 620, Sa Lu Zi carried Taizong, who was then a prince, to vanquish Wang Shicong in Luoyang, Henan. In a dashing move to penetrate into the enemy's rear, Taizong ended up confronting the enemy on his own while riding Sa Lu Zi. Unluckily, a stray arrow struck the horse, but Qiu Hanggong, a valiant guard whose horsemanship and marksmanship were excellent, came to the rescue at the crucial moment. Qiu gave his horse to Taizong and then yanked the arrow shaft out of Sa Lu Zi. While holding Sa Lu Zi with one hand, Qiu used his sword to fight the enemy with the other. In the midst of a melee, Taizong and Qiu killed a number of the enemy and raced back to their barracks. Sadly, Sa Lu Zi died from its wound soon after.

The plaque depicts the poignant moment when, shot by the arrow, Sa Lu Zi appears in pain with its head held low; Qiu meanwhile, with his curly beard and war cloak, is using one hand to caress the horse and the other to pull out the arrow shaft. The intense affection towards the horse, and the symbiosis of rider and horse, are startlingly vivid and stir the human soul.

The second plaque on the west wall was that of Quan Mao Gua (Fig. 2.11). "Gua" in Chinese means a yellow (champagne-coloured) horse with a black muzzle. "Quan Mao" means curly hair. It was common currency at the time that a horse with curly hair was inferior and ugly. Nevertheless, Quan Mao Gua's impressive speed and tenacity made Taizong choose it as his charger in spite of its less than perfect appearance.

Quan Mao Gua is seen as having been hit by three arrows in its front quarters and six in the back in the battle to quell Liu Heita. It ultimately died from these wounds.

The last plaque on the west veranda was Bai Ti Wu (Fig. 2.12). "Bai Ti" in Chinese means white hoof and "Wu" means black. It was thus a black horse with four white hooves. Bai Ti Wu carried the emperor to fight Xue Rengao.

SIX STEEDS OF ZHAOLING MAUSOLEUM
ZHAO LIN (ACTIVE IN THE 12TH CENTURY)
HAND-SCROLL, 27.4 x 444.2 CM
PALACE MUSEUM, BEIJING
(FIG. 2.13)

Zhao Lin's painting was modelled on the original stone reliefs of *Six Steeds of Zhaoling Mausoleum*.

Fig. 2.13 Six Steeds of Zhaoling Mausoleum (Hand-scroll).

This is Zhao Lin's only extant work and includes the same inscriptions of Emperor Taizong's eulogies by the renowned Jin empire (1115–1234) poet and calligrapher Zhao Bingwen (1159–1232). The painting is reminiscent of the Tang style, especially that of Master Han Gan (706–783) and the Northern Song School.

Using meticulous and precise brushstrokes, heavy colour, line drawing and colour washes alongside close attention to postures, riding tack and facial expressions, the painting is both visually arresting and dramatically effective, freshly capturing the many virtues of the carvings.

During the Song dynasty, north-west China was occupied by the alien Jin empire. Replication of the Tang plaques mirrors Zhao Lin's patriotism and his longing for the Tang dynasty.

Nevertheless, small discrepancies exist between the plaques and the painting. The tack set in each artefact exhibits the style of different eras. The cylinder-shaped quiver in the plaque of Sa Lu Zi is different from the tiger-skin quiver in the corresponding figure in the painting. Also, the plaques depict the horses as "three-flower" horses with neatly docked manes fashioned in three crenellated points. The painting does not. This style of mane braiding was brought to China by the Turkic peoples and was preserved only for royal horses. Equestrian artefacts at the time of the Tang were commonly modelled on three-flower horses.

Night Shining White
Han Gan (706–783). 30.8 x 34 cm
Metropolitan Museum of Art, New York
(Fig. 2.14)

Han Gan is one of the renowned masters of Chinese equestrian art. He was a court painter during Emperor Xuanzong's reign (712–756) and is best known for his astonishing comprehension of equestrian anatomy and for grappling with the concept of a real horse in motion.

Emperor Xuanzong had two favourite blood-sweating horses, both of which were presents from Fergana when Princess Heyi married a royal of the home country of blood-sweating horses. Xuanzong gave the two horses the names of Zhaoyebai (Night Shining White) and Yuhuacong (Jade Flower Grey). Han Gan's portrait of Night Shining White is one of the most famous representative equestrian paintings in Chinese art. The unfettered steed is portrayed with remarkable energy and strength.

With a rotund torso and short skinny legs, the imperious horse strives for freedom – stamping its hoofs, throwing up its head, widening its nostrils and neighing. Alas, to no avail: the horse is tethered to a thick iron pole. Alluding to the tragedy of court life for a horse, the animal is personified with its agonized eyes as appealing for sympathy. The expression on the horse's face is almost human in its desire and despair, emotions which nevertheless are neither devoid of dignity nor inconsistent with its pride.

The horse appears far from a realistic rendition of a Central Asian steed. Du Fu criticized Han Gan for "merely depicting the horse's flesh but not its bones". Actually, as Zhang Yanyuan said in his defence, Han Gan was instead "capturing the horse's essential spirit". The portrait combines both a realistic rendition and the subjective image in the painter's mind. The horse was the emperor's best-loved steed; it was well fed and ridden only by the emperor. Whichever way it is viewed, it looks plump. In fact, the animal's clumsy physique complements and accentuates its rebellious spirit.

Fig. 2.14 Night Shining White.

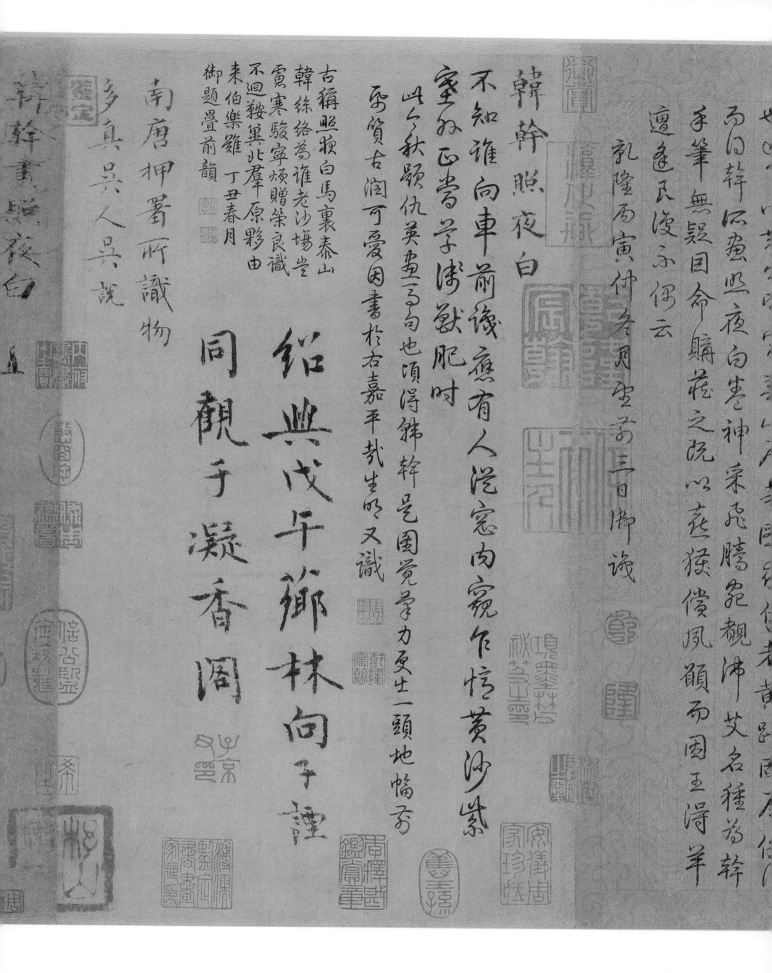

SILVER SADDLE-FLASK WITH
GILDED DANCING HORSES
18.5 CM HIGH
SHAANXI HISTORY MUSEUM, XI'AN
(FIG. 2.15)

This silver wine flask with its gilded dancing horses was unearthed in October 1970 among a hoard of some 270 pieces of silver and gold wares at Hejia Village in the southern suburbs of Xi'an, the capital of the Tang dynasty.

Shaped in the style of a nomad's leather bottle, or as some see it resembling a stirrup, it is flat and round, and raised on a foot with a beaded rim. The lotus-petal-shaped spout-stopper is attached by a chain to a handle. Each side of the flask has a dancing horse in repoussé relief that appears plump and docile with a tidy and smoothly arranged mane hanging down the forehead and neck. The horses' tails are flicked upwards, while the ends of the knotted ribbons flap in the wind. Finally, the two engraved horses are portrayed squatting on their hind legs, as if bowing, while each clenches a wine cup in its mouth. This exquisite treasure attests to the grand scene of horse dancing during the birthday celebration party of Tang dynasty's Emperor Xuanzong. Before the discovery of this flask, horse dancing (see the Equestrian Sports chapter) was only recorded in ancient literary texts, from which it is difficult to visualize the legendary choreography. The flask, which demonstrates the feat, depicts the last scene of the most acclaimed music for horse dancing, the *Song of the Upturned Cup*. In a lively fashion, the horse drops its head, flicks its tail, bends its hind knees and clenches a wine cup in its mouth as if to toast the emperor.

This piece also provides evidence of the cultural exchange between nomads and the Chinese. The saddle-flask was a popular shape for the period, although ceramic flasks are the more common.

The wine flask illustrated here is the oldest of any similar excavated artefact.

The extremely high standard of the craftsman's technique is very clear. The surface is rounded and silky smooth without any apparent seam. The concavo-convex horse relief on the surface was made by first creating a three-dimensional mould and then very precisely hammering from the inner part of the flask to create the contours and muscle lines of these two beautifully executed gilded dancing animals.

Fig. 2.15 Silver Saddle-flask with Gilded Dancing Horses.

TRI-COLOURED GLAZED POTTERY HORSES

Tang tri-coloured pottery (Tang *sancai*) is renowned for its richness of colour, vibrancy, exuberance and exquisite detail, and is so called because of the liberal use of yellow, brown and green as the basic glaze colours. The craft of glazing was developed back in the Han dynasty and reached its apogee in the Tang dynasty.

The motifs used in Tang tri-coloured ceramics range from the human and animal form to vases, and of these the horse is among the most eye-catching.

Artisans primarily depicted fine horses, which are believed to have been imported from the west, featuring strong necks, aquiline heads, fleshy and well-defined bodies. They come complete with ornate tack, docked tails and all-over rich glaze. Their corpulence, which is in keeping with the image of the Tang court ladies, was the vogue of the time.

Very rarely found is the tri-coloured horse in motion. Instead, the horse is almost always sculpted standing four-square, with its face projecting forwards (Fig. 2.16), casting to the left or whinnying. The dynamism of the animal is nevertheless encapsulated through the rolling and lively eyes and pricked-up ears, while its spirit is sympathetically and subtly captured.

Fig. 2.16 Tri-coloured glazed pottery horse. This beautiful example of a "three-flower" sancai-glazed horse shows its neatly docked mane. This braiding style was preserved only for royal horses, making three-flower horses popular subjects for Tang artefacts.

Fig. 2.17 Ethnic minority groom.

Fig. 2.18 Woman polo player.

Fig. 2.19 Five Tribute Horses and Grooms. This painting had disappeared in China for almost 90 years and was found displayed in the Tokyo National Museum, Japan, in early 2019. The inscriptions by Huang Tingjian, a noted Song dynasty Chinese calligrapher, are of very high artistic value. Interestingly, the order of the horses in all the extant collotypes varies from this original painting.

Pottery horse sculptures are most often accompanied by a rider or a groom who is crafted as a foreigner or a member of a minority group (Fig. 2.17).

It is also very common to find women rider figures (Fig. 2.18). Against the backdrop of the Li imperial family's nomadic lineage, which resulted from the intermingling with steppe nomads over several generations, women of the Tang dynasty were blessed to have lived at a time of broad-mindedness and tolerance. They enjoyed a higher status compared with those of other dynasties during China's feudal period.

Tri-coloured pottery was not used for household utensils as it contained lead. Of all the tri-coloured ceramics employed as burial articles by the gentry and other wealthy people, horses are the most commonly excavated figures. This once again attests to the Tang people's fondness for these animals.

The precise and loving details put into the artefacts also speak of the social life of the Tang dynasty and the extensive cultural exchange during its peak when merchants, diplomats and missionaries from all over the world gathered in the capital Chang'an and the business centre Louyang. By gazing on these exquisite wares, we can transcend time and space to gain an insight into the flourishing pre-eminent civilization of the Tang dynasty.

The period in which tri-coloured glazed pottery was in demand witnessed the coming to power and prosperity of the Tang dynasty. The Tang's lavish burials were begun during the reign of Emperor Tang Gaozong (reigned 649–683), continued during the reign of Emperor Tang Xuanzong and then started to decline after the rebellion of An Lushan (755–763), when society began to reject the extravagant lifestyle of the dynasty's former years and heralded its downfall.

SONG DYNASTY (960–1279)

FIVE TRIBUTE HORSES AND GROOMS
LI GONGLIN (1049–1106). 27.8 x 256.5 CM
TOKYO NATIONAL MUSEUM, TOKYO
(FIG. 2.19)

In the history of Chinese art, the Song dynasty produced the least number of equestrian artefacts because the Song court considered literacy more important than military endeavours. Li Gonglin was one of the very few painters who specialized in horses during that period.

Li Gonglin was born into a scholar-gentry family and became an officer. Despite his lacklustre public career, he is well known for his portraiture of horses, and Buddhist and Taoist figures.

Five Tribute Horses and Grooms vanished in China in the waning years of the Qing dynasty, taken away

from the Forbidden City and feared destroyed. However, it reappeared in the "Unrivalled Calligraphy: Yan Zhenqing and His Legacy" exhibition held in the Tokyo National Museum in January and February 2019. The painting was shown as a side exhibit to downplay its sensitive provenance, but in fact the masterpiece emphatically stunned Chinese art enthusiasts around the globe. The painting had been kept in Japan for almost 90 years in a private collection and recently had been donated to the Tokyo National Museum, the owner wanting all those concerned about this chef-d'oeuvre to be reassured of its safety and excellent condition.

According to the sales document, the painting was sold by Liu Xiangye to art collector Suenobu Michinari through the antiques dealer Eto Namio on October 1st 1930. It seems Liu Xiangye, whose uncle was Chen Baochen, a teacher to the last Chinese emperor, Puyi, sold many national treasures at around that time to support his hedonistic and lavish lifestyle.

Li Gonglin's fame in Japan began in the Kamakura period (1185–1333), and the painting is now highly valued in the country and classified as an important art object. Interestingly, while there are several collotype copies in circulation, we can now see that the order of their horses varies from the original painting.

The evocative hand-scroll comprises the separate paintings of five tribute horses which were presented to the Song court between 1086 and 1089, each attended by a foreign or Chinese groom. They have been spliced together to form the scroll and are accompanied by several inscriptions. Huang Tingjian, the finest calligrapher of the Song dynasty, wrote the most important inscriptions noting the name, time of tribute, place of origin, stable name, age and size of each of the first four horses, as well as confirming Li Gonglin as the artist. The last portraiture does not contain any of these and the technique and style, as well as the ink and paper texture, differ from the other four; the painter's identity is thus unconfirmed. Emperor Qianlong of the Qin dynasty and another Song calligrapher, Zeng Yu, each added further inscriptions to the painting.

The five horses from right to left are Fengtoucong, Jinbocong, Haotouchi, Zhaoyebai and Manchuanhua. Purposefully reduced to finely controlled and supple ink outline without the decorative appeal of setting, the painting visually conveys sharp distinctions of character. Viewers can thus tell from afar that the horses, which appear top-class, poised and tractable, are all tribute horses from the west, from places like Khotan and Tibet.

Close attention is paid to the peculiar bonds of sympathy the painter perceived between the horses and their grooms as each pair tallies uncannily well through tiny but telling details. Fengtoucong and its groom both reveal a magnanimous and calm disposition. The green and timid Jinbocong is led by a composed groom with a droopy hat.

Not until the masterpiece resurfaced was it realized that the scroll is coloured rather than being monochrome. Furthermore, as fluorescence and near-infrared photography reveal, a wide range of hues has been used whose application is very complicated and highly sophisticated, beyond our previous understanding of Northern Song painting.

Glazing washes are applied to the horses' bodies in a variety of tints of different values to enhance the perception of three dimensionality. The rosy cheeks of each of the first two grooms show the blending of ink and assorted pigments to create the illusion of depth and space and to accentuate the configuration. The first foreign groom's chestnut beard and moustache are overlaid with a mix of distinct and blurred colours, this adds to the subtlety of the work and gives it a more polished and finished look.

In China, this artwork has been revered since its completion in the late 11th century. Accounts of the painting in several historical sources vary from the original on things such as the order of the horses and its inscriptions. While Japanese art experts, led by Itakura Masaaki, have debunked many mysteries about the masterpiece, including how it got from China to Japan, its composition, the techniques and papers used, the painter's identity and so forth, many other conundrums are still waiting to be answered. It has been an exciting find.

EMACIATED HORSE
GONG KAI (1222–1307). 29.9 x 56.9 CM
OSAKA CITY MUSEUM OF FINE ARTS, OSAKA
(FIG. 2.20)

Gong Kai held a minor office in the late Song dynasty. When the Mongols overthrew the Song and established the Yuan dynasty, he had no job and led a harsh life, making a meagre living as a scholar-amateur painter, selling paintings and calligraphies. He is notable as the first symbolist painter in the Yuan, and *Emaciated Horse* is his only extant work of a horse, and his most famous painting.

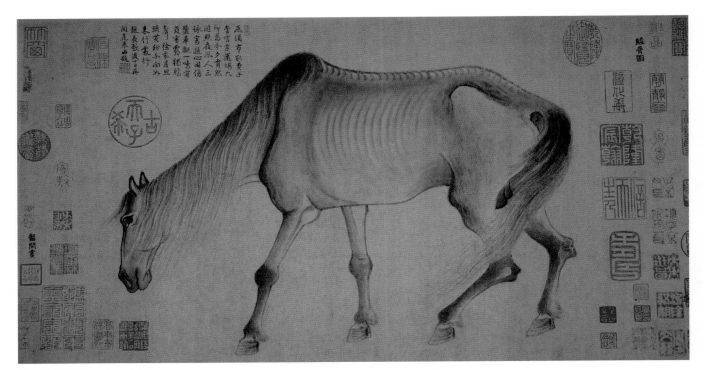

Fig. 2.20 Emaciated Horse.

Fig. 2.21 Mounted Official.

The scrawny but still noble animal is seen as the embodiment of the Song loyalist who was now living under an alien regime. The 15 visibly projecting ribs demonstrate the horse's steadfastness in the face of adversity. The sickly animal drops its head, its mane falling low in misery and distress, yet the eyes are still strong. The whole image is permeated with an ambience of infinite glumness that comes, nonetheless, tinged with a spirit of patriotism fuelled by former times.

Gong Kai was to influence other artists during and after his lifetime whose work was a symbolic protest against Mongol rule. Undoubtedly, this painting spawned the appearance of several emaciated horses, associating them with loyalty and integrity. The lean horse in *Fat and Lean Horses* by Ren Renfa (page 58) is an example.

Gong Kai also composed a poem and inscribed it on the left-hand colophon section of the hand-scroll (not shown in Fig. 2.20). It reads:

> "A celestial horse passed the heavenly gate through clouds and mists to the earth, completely outshining royal steeds in the twelve stables of the previous dynasty.

> Today, who will sympathize with this shrunken horse, its mountain-like sunset shadow casting on a sandy shore?"

The celestial horse in the poem symbolizes the painter who when young capably served the Song court. Although now he is old and frail, his constancy is still as solid as a mountain.

YUAN DYNASTY (1279–1368)

MOUNTED OFFICIAL
ZHAO MENGFU (1254–1322). 30 x 52 CM
PALACE MUSEUM, BEIJING
(FIG. 2.21)

Zhao Mengfu was the 11th great-grandson of the founding emperor of the Song dynasty but became an official under Kublai Khan in the Yuan dynasty. Although Zhao was a senior officer in the alien dynasty, he neither had much official authority nor any opportunity to participate in important issues. He was so frustrated that he diverted himself with painting, poetry and calligraphy.

The literati painter had a particular aptitude for depicting horses, painting for personal interest and pleasure. Mongol aristocrats and officials notably loved horses and Zhao also painted a vast number of equestrian hand-scrolls for them.

The painting *Mounted Official* depicts a well-poised and calm officer in a red gown, riding a plump and sturdy horse at the walk. The red gown resembles a Tang official uniform and the equine is an oriental horse rather than a Mongolian horse, which was commonly seen in the Yuan dynasty. In addition to Zhao's imitation of Han Gan's Tang style, the painting reflects his nostalgia for this most glittering period in China's history.

Zhao had dated and written on the painting the title as well as a short colophon: "I have loved painting horses since childhood. Recently, I had the chance to see three authentic scrolls by Han Gan and started to understand the meaning behind them."

Emperor Qianlong of the later Qing dynasty favoured the painting very much and added it to the royal collection. He affixed 16 royal seals to the painting and inscribed a poem he had composed: "It is difficult to find a premium horse while it is more difficult to capitalize on its talent. Zhao painted this hand-scroll in his free time, which brought the connotation to my mind." Qianlong made an analogy between a quality horse and a capable officer. As an emperor, he had his own unique interpretation of the painting.

FAT AND LEAN HORSES
REN RENFA (1255–1328). 28.8 x 142.7 CM
PALACE MUSEUM, BEIJING
(FIG. 2.22)

Ren Renfa was not a court painter. He first served in the Mongol Empire as a hydraulic engineering official but was later promoted to an assistant controller for irrigation. Although he had served the alien dynasty throughout his working life, he preserved his conscience, dignity and integrity. He sought to express his own inner voice through his paintings, revealing and satirizing the rotten side of society and the less desirable of human traits.

This painting is executed as an allegory, drawing an analogy between good and bad to criticize the malfeasance of the day. The muscular skewbald horse, with its swaying tail, carries a rein which is trailing along the ground, while raising its head and

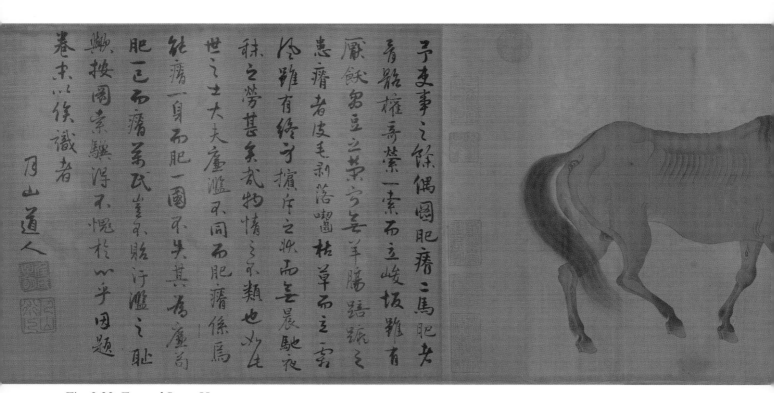

Fig. 2.22 Fat and Lean Horses.

swaggering perkily. It represents a corrupt official who takes advantage of his position to enhance personal interest. The loose rein signifies that the official goes unhindered in fuelling his personal concerns as a result of lax controls.

The lean chestnut horse, on the other hand, with its limp tail, carries a rein which is secured round its neck, while it droops its head and shuffles listlessly. This horse represents a righteous official who scorns personal interest for the welfare of the people. The rein held firmly round the horse's neck betokens that the good official performs his duties with fidelity and integrity under appropriate supervision.

QING DYNASTY (1644–1911)

ONE HUNDRED HORSES
GIUSEPPE CASTIGLIONE (1688–1766). 94.5 x 776.2 CM
NATIONAL PALACE MUSEUM, TAIPEI
(FIG. 2.23)

Giuseppe Castiglione was an Italian Jesuit missionary who had been born in Milan. In 1715, the talented Western-trained artist made the long journey to Beijing to assist the missionary work of the Jesuits through acting as a court painter and architect over the reigns of Kangxi, Yongzheng and Qianlong, the three greatest emperors of the Qing dynasty. He was also known by his Chinese name, Lang Shining.

Despite the considerable honour bestowed on him as a foreign court painter and a confidant of the three emperors, Giuseppe Castiglione had limited influence on court life. Kangxi was of the opinion that Christianity was against traditional Chinese thinking and did not give Giuseppe much chance to preach the gospel. Instead, he spent his life toiling away year after year in the palace's enamelling workshop from seven in the morning till five in the evening painting what and how he was told. He had to set aside the painting techniques and lore he had learnt in Italy, adopt a wealth of new artistic values and follow a string of rules. Among other things, shading was not allowed in portraits as Emperor Qianlong considered the technique made his face look dirty. Every painting had to be drafted for the emperor's approval before execution and there

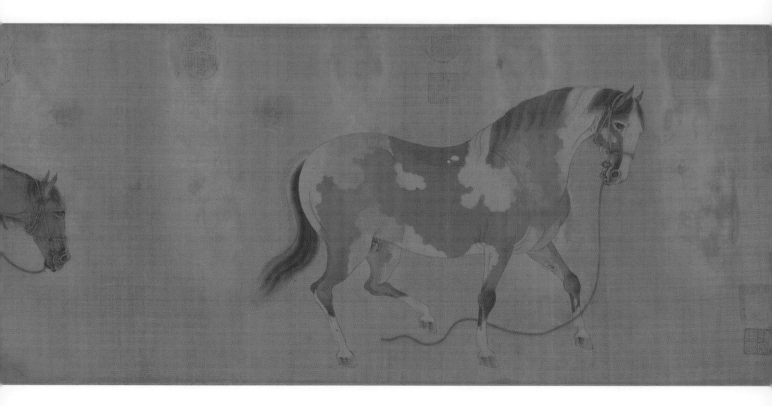

Fig. 2.23 One Hundred Horses.

thus exists the preparatory drawing of *One Hundred Horses*, which is now kept in the Metropolitan Museum of Art, New York, USA (Figs. 2.24–2.26).

Having said that, Giuseppe introduced into Chinese art culture such techniques of Western virtuosity as atmospheric perspective, three dimensionality and shading. These techniques are derived from geometry, which can help make precise measurements of light and shade, foreground and background, conveying a more realistic and sprightly illusion. A new, synthetic style emerged in which Western realism permeated and enhanced traditional Chinese conventions of composition and brushwork.

The horse was the favourite animal of emperors and to nobody's surprise the animal is a major motif of court painting. *One Hundred Horses* is a long hand-scroll which delineates this age-old subject of Chinese painting through Western techniques.

Fig. 2.24 Detail from Giuseppe Castiglioni's preparatory drawing shows a distinctly emaciated horse.

Fig. 2.25 Hind-view of a second emaciated horse.

Fig. 2.26 A third emaciated horse stands listlessly.

Giuseppe was commissioned to paint his *One Hundred Horses* in the year of Yongzheng's accession to the throne. The eight-metre hand-scroll took him five years to complete and is considered his crowning glory. However, for some reason, Yongzheng never saw the scroll. Eighteen years later in 1735, it was presented to the newly enthroned Emperor Qianlong who hailed it as a masterpiece and appointed Castiglione his chief court painter. Later at Qianlong's behest, he sketched another preparatory drawing that in reverse was based on the finished silk scroll. However, Giuseppe was asked to draw only the horses while two Chinese court painters were to depict the grooms, trees, rocks and mountains. This arrangement arose because of Castiglione's over-realistic rendition of the objects in the original hand-scroll, which resulted in the loss of the traditional Chinese atmosphere — something that was highly valued by the emperor. Qianlong was fond of inscribing on artefacts but on this occasion he added a poetic inscription only to the paper scroll. The silk scroll appears to have been ignored by the emperor, but as a consequence, it remains clean and intact as an artefact.

Yongzheng and Qianlong both took art seriously. Nevertheless, Yongzheng allowed his artists more artistic freedom of expression while Qianlong was eager to participate in the art's creation himself, especially in work allocation. During the reign of Qianlong, Castiglione had fewer opportunities for his own artistic creativity, often having to cooperate with Chinese painters when he was assigned to paint only animals.

In the hand-scroll, the horses are rendered in a range of actions from grazing, rolling, grooming, nickering and neighing to ambling, snorting, crouching and sleeping. The landscape is drawn on a Western-style perspective, figures are often shown using the practice of foreshortening and vegetation is represented using cross-hatching. The multi-point perspective of Chinese painting is also applied. The pine trees, which serve to divide the composition into discrete and yet continuous components, each with its own vanishing point, allow viewers to move their eyes along the scroll without being confined to one fixed static point of view. It is a strong execution of the syncretic style of East and West.

On close examination, viewers can find three emaciated horses (Figs. 2.24–2.26) hidden among the many other strong, energetic and handsomely noble horses. Their skeletal frames are clearly depicted to arouse viewers' sympathy and reflect Castiglione's masterly comprehension of the anatomy of the horse.

Horses symbolize talent in Chinese culture and the hand-scroll implies that the Qing court was full of capable officials. On the face of it, the painting demonstrates a very serene and auspicious atmosphere. Castiglione however really associated

Fig. 2.27 Galloping Horse.

himself with the three less happy horses, finding that they echoed his solitariness.

Emaciated horses are also to be found in his other works. These include *Eight Prized Steeds* and *Horse Herding in the Suburban Field*. Cao Tiancheng, an art historian, suggests that their presence was not unintentional.

Castiglione was alone in China. It was a foreign country, he was not allowed to spread Christianity and his artistic freedom of expression was constrained. He was frustrated in all his endeavours. The expression of his innermost feelings at such long-term obstruction is projected through the symbolism in his work.

MODERN CHINA (1911–PRESENT)

GALLOPING HORSE
XU BEIHONG (1895–1953). 130 x 76 CM
XU BEIHONG MEMORIAL MUSEUM, BEIJING
(FIG. 2.27)

Xu Beihong was born into a poor painter's family in Yixing, Jiangsu. He had learnt traditional Chinese painting from his father from the age of six. Supported by friends and a scholarship, Xu had the chance to study in Japan and France, and embarked on a quest for artistic modernity. As an artist living at a time of change and war, Xu was profoundly influenced by the cultural leaders of the May Fourth Movement and thus sought to uphold the principle of "depicting reality". His fusion of Western realism and traditional Chinese brushstroke made him one of China's most forward-looking and highly acclaimed artists. He is the most important and influential shaping force in modern Chinese art, as well as a source of inspiration for many of his contemporaries and even of painters today.

The master of horse painting in modern China had nursed a lifelong fascination with the quadruped. To better his prodigious talent, he frequented the outlying countryside to research the animal's physique, behaviour and movement, studying its anatomy to understand its muscles and skeleton, and most of all, making innumerable sketches. Eventually, he painted from the inside out and could depict a horse without looking at the animal or a drawing of it. Attentive to detail and subtle nuances, he preserved the physical and sensuous beauty of the animal. His horses are almost always free of tack and rider, and the wild and untamed image that results embodies his appreciation of freedom and desire for liberation. The historical painting *Jiufang Gao* is one of the few exceptions where a horse is tied with a halter (Fig. 2.28). *Jiufang Gao* was created in 1931, narrating the story of the genius horse judge Jiufang Gao. The black mare wearing a halter in the painting is a prize horse. Xu explained that horses are just like humans, willing to serve the person who appreciates them but reluctant to submit to an inept leader.

He persisted in Western academic realism while executing his work with Chinese ink wash and line. The horse under his avant-garde style exudes an air of unique and fascinating allure.

During the early 20th century when China was being bullied and exploited by colonial powers, Xu was so overwhelmed with emotion that he made his paintings a vehicle to express his patriotic fervour and to help rouse the Chinese to resist foreign aggression.

Xu Beihong's *Galloping Horse* was shown in autumn 1941 when he staged an exhibition in Penang, Malaysia to raise funds for the War of Resistance against Japan. Worried about the second Changsha battle, Xu conveyed his feelings through this spontaneous painting, completing it overnight. Created as a metaphor, *Galloping Horse* is a patriotic soldier, symbolizing valour and prepotency. The ink painting weaves a vision of a galloping horse from the Romantic period with sentiments of patriotism; it is an image of agitation combined with ebullience. The horse is represented in perspective, with the larger forelegs and head and scaled-down hindquarters.

SEEING BEYOND THE IMAGES

Throughout China's long and complex history, the horses created under the influence of the country's distinctive artistic and cultural traditions have been given a peculiar image and aura of their own. Artistically, they may look somewhat identical,

Fig. 2.28 Xu Beihong's horses are almost always free of tack and riders, except in the historical painting *Jiufang Gao* where a horse is held by its halter.

produced with less detail and less diversity than their Western counterparts, but these artistic feats are often a symbol for much beyond the physical body of the artefacts.

Chinese artists not only present the illusion of the horse with scrupulous accuracy but also the image in their mind's eye. Unlike most Central Asian steeds, *Night Shining White* is depicted with a rotund body but scraggy limbs. This grotesque image is more or less intuitive rather than being purely aesthetic or realistic. It reveals the background and emotions of the emperor's pet. Seeing beyond the image to the connotations involved allows viewers to imagine how the story unfolds or even triggers their imagination, which is something that goes well beyond the artist's intention. Other examples are *Emaciated Horse and Fat and Lean Horses*. They are both made to be metaphorically relevant, carrying moral and ethical messages.

Smothered with allusions and references, Chinese equestrian artefacts always invite viewers to call on ancient Chinese history and literature as well as the doctrines of Confucianism, Taoism and Buddhism. The artistically important horse sculptures in the tomb of Huo Qubing tell us

of the heroic actions of the Han general; *Six Steeds of Zhaoling Mausoleum* reminds us of the accomplishment of Emperor Taizong of the Tang dynasty. *Galloping Horse* echoes Confucianism's loyalty and patriotism while *Bronze Cantering Horse* resonates with Taoism's carefree and ethereal spirit.

WHERE ARE THE RIDERS?

Strangely, we see very few riders in Chinese equestrian art. According to Wang Shen of Nanjing University of the Arts, this is a reflection of the spiritual insights of Taoist practice.

Mother Nature is the crown jewel in Taoism. The goal of Taoists is to attain harmony with the Tao: in other words, to live in accord with nature. The notable absence of rider suggests that humans are nothing but a small part of nature and can be disregarded in the context of Taoist aesthetics. Having said that, being and non-being are the same and they produce each other. The absence of rider is just a feeling, an artefact of cognition instead of an actual absence. On the other hand, the presence of a rider is anything but an actual rider, which too is an artefact of cognition. The Tao is the root of cognition.

Perhaps if Taoism is too metaphysical to grapple with, we can comprehend the absence of a rider from a secular perspective.

Chinese artists loved to pick horses from the west rather than indigenous breeds. These fine horses were all precious to China in terms of their military capability and fascinating appearance. Meanwhile, the riders were few and far between, mainly nomads, ethnic minorities and members of the gentry, who, like the horse, were out of reach of most ordinary people, including the Chinese artists. The more people were unfamiliar with something, the more curious they were about it. The mythical and sublime background of the horse led to the animal being associated with dragons and centaurs in ancient literature and legends, thus generating a celestial status.

WHO WERE THE ARTISANS?

However, the subject of this artistry represents only half of the story of the horse in Chinese art; the other half belongs to the creators of these outstanding works.

In Western societies, artisans were highly respected. Prominent examples are Michelangelo and Leonardo da Vinci, both being well acquainted with the upper class and enjoying enormous prestige and fame for centuries by virtue of their prodigious masterpieces. However, while China produced a huge volume of world-class treasures, one is hard put to name any world-renowned Chinese artisans. In China's structured feudal society, painters were considered as having a lowly occupation, while artisans were even worse. Painters could be officials and literati but artisans most often were illiterate or poorly educated. Their work was mainly commissioned by rulers or officials and their creativity could not develop to its full potential.

Yan Liben, the consummate painter of *Six Steeds of Zhaoling Mausoleum* mentioned earlier, was a prime minister in the Tang dynasty. Before ascending to the highest rank in the cabinet, he was commissioned to paint for the court from time to time. He felt ashamed at being called a "painter" and, unsurprisingly, frowned upon his children learning to paint.

Guo Xi was a renowned painter in the Song dynasty. His son who was a court official felt so embarrassed about his father being a painter that he bought back all of his work at high prices in order to hide his inferior provenance.

As the status of artisans in ancient China was even lower, they were seldom mentioned in historical documents; and if they were mentioned, details of them were not recorded.

Set against the rousing equine subject brimming with both mystique and majestic spirituality, the humble artisan painters were so awed that they barely imagined conquering or dominating. As a result, the animal was almost always depicted without saddle, bridle and rider – as unfettered as it had been created at the beginning of history.

What is also significant, Western achievement-based society encourages individualism that projects each individual as an integral part of the universe and society. Chinese culture gears towards collectivism, which makes an individual just a part of the group and subservient to the goals of the group. Accordingly, rather than through the riders, symbolism through the horse is emphasized.

Man together with the horse form a perfect symbiosis in Western equestrian artefacts. The animal's stately bearing and exquisite appearance help to enhance the reputation and projection of those on its back. Classical examples include Napoleon, Apollo, El Cid and Queen Victoria. Chinese emperors on the other hand chose to use such symbolisms as cloud, sun, moon and dragons to represent their sons of heaven status. As the authoritative emperors did not make the most of their steeds to convey their nobility, heroism, divination and idolization, there was no way that anyone else could mount the magnificent quadruped in the realm of art.

CHAPTER THREE

EQUESTRIAN SPORTS

INTRODUCTION

In ancient China there was no such term as sport. Vibrant and diverse recreational activities however did exist and those of an equestrian nature were among the most impressive. This period of imperial history spawned a wealth of equestrian pursuits, and in accordance with the social and cultural conditions at each historical stage, they epitomized the distinctive spirit and features of the times.

Equestrian activities first evolved from hunting; they then emerged in military endeavours and eventually spread into the civil community as physical exercise and leisure pursuits. It is believed that nomads in northern China first developed horsemanship. Inner Mongolia, Ningxia, Qinghai, Gansu and Xinjiang are the former homelands of nomads like the Turkics, Xiongnu, Da Rouzhi, Xianbei and Qiang. Along the region, great numbers of rock carvings which date back to between 4,000 and 10,000 years ago depict scenes of hunting and fighting on horseback (Fig. 3.1). These are all relics of ancient nomads.

Fig. 3.1 Prehistoric rock carving from the Helan Mountains, Inner Mongolia, shows hunting on horseback.

PHILOSOPHY OF EQUESTRIAN PURSUITS

The two most influential philosophical schools, Confucianism, from 500 B.C., and Taoism, from 400 B.C., both support physical exercise. The great philosopher and educationist Confucius (551–479 B.C.) himself was fond of archery, fishing, hunting, walking and hill-climbing, and he had a particular interest in charioteering. His philosophy of education placed an emphasis on a person's moral, intellectual and physical development, and he espoused an education system comprising six arts: charioteering, rites, music, archery, calligraphy and mathematics, which he claimed complemented each other. For example, charioteering was governed by a complex system of rules and etiquette, underlining the cultivation of virtue, self-improvement and recreational pleasure, a concept that is comparable with "sportsmanship" today.

Taoism underscores longevity through various techniques of breathing, diet and exercise sequences known as *daoyin*. These exercises are an ancient precursor of qigong and an imitation of animal actions.

Muscle development, which has been highly valued in modern Western culture, was however not appreciated by Taoism and Confucianism.

Sports in Chinese civilization were more about virtues and practicality, encouraging harmony and entertainment. Ancient China was rooted in its non-itinerant agricultural way of life, in which competitive practices were politically and commercially estranged, giving rise to the development of non-competitive forms of recreation. As such, physical activity focused on the process instead of the outcome, while players sought to prove that recreation could bring out the best in themselves. As emulation, strife and rivalry were likely to damage the harmonious patriarchal social structure, which was of upmost importance to the ancient Chinese system, there was little room for competition.

It is argued that polo as a form of military training was intensely competitive in nature as players attacked and defended their goals. The underlying precept should in reality be looked at from a different perspective. Rather than its purpose being to defeat an opponent, the competition was more a test of an individual's ability to nurture and improve his own skill and virtue while following all the correct rites. Simply put, extrinsic rewards and victory were marginalized. In a polo match, players challenged themselves in such a way as to accomplish a harmonious whole.

This belief in the value of sport resonates with the modern Olympic creed, whose guiding principle is: "The most important thing in the Olympic Games is not to win but to take part, just as the most important thing in life is not the triumph, but the struggle. The essential thing is not to have conquered, but to have fought well."

Despite the passage of time, the spirit of sport remains unaltered although rules and conditions have changed in consequence of social development.

NOMADIC ORIGINS

The variety of ethnic groups within China, their geographical location and traditional occupations used to determine who took part in equestrian sports. The nomadic peoples led a lifestyle which required life skills like horse training, hunting and horseback exercises that were basically warlike

Fig. 3.2 Silver Lining was a legendary horse of the late 1970s in Hong Kong. It won the Champions and Chater Cup three times and was three times voted Horse of the Year, in 1978, 1979 and 1981.

habits. In contrast, the settled cultures along the coastal plain were guided by philosophical directions which tended to be less competitive in their approach to life and recreation. These people practised gentler forms of recreation such as archery, daoyin and football, and they stressed the importance of cooperation and harmony at the expense of competition. The native environment of the nomadic peoples was the harsh steppe and semi-desert, which are more favourable to horse riding than the central plain, where the settled agrarian Chinese communities lived.

HORSE RIDING IN CHINA TODAY

Nowadays, equestrian pursuits are no longer constrained by such intrinsic factors. There are racecourses in Hong Kong (Fig. 3.2) and Macau, neither of which have a pastoral heritage or a habitat that is natural to horses. Hong Kong and Macau are on the equatorial boundary of the subtropical climate zone characterized by hot, humid summers and cool winters.

Fig. 3.3 A show jumping fence engraved with Chinese dragons at the 2008 Olympic Equestrian Events in Hong Kong.

Fig. 3.4 Dressage arena with Chinese pavilions for judges and officials. Olympics 2008, Hong Kong.

The summer is oppressive with temperatures over 30°C and an average humidity of over 80 per cent. The Hong Kong Jockey Club does not hold any race meetings during the height of the summer, starting them again in early September, which is sometimes still muggy enough to get the horses into trouble. In the 700 races staged each season, about four or five horses manifest unsteadiness or other heat-related problems. Once every season or two, a horse collapses from heat exhaustion, but it is quickly revived and does not develop long-term health problems from the experience.

The 2008 Olympics were held in Beijing but the equestrian events were held in Hong Kong (Figs.

3.3 and 3.4) as Beijing failed to establish an internationally recognized equine disease-free zone. If the equestrian events had taken place in Beijing, the horses would have had to endure lengthy periods in quarantine upon returning to their home countries.

Hong Kong is one of the most tropical places in which an Olympic event has been held; it has a much higher humidity on summer afternoons than sultry Atlanta, which hosted the 1996 Olympics, and hitherto had been the toughest climate with which Olympic equestrian teams had had to contend.

The humidity poses a particular challenge for the event because the horse is the only mammal, other than humans, that cools itself by sweating.

When the humidity is high, the rate of evaporation of sweat from the skin decreases, resulting in a body temperature that is much higher than the ambient air temperature. Worse, the oppressive heat can make the horses susceptible to hyperthermia, or overheating, and this in turn can lead to death if not treated properly and quickly. It can be a crucial factor in the intense and strenuous cross-country test.

Although horses tend to manage better in hot and dry conditions, problems presented themselves at the 1968 Olympics, which were held slightly closer to the equator than Hong Kong in Mexico City. Lying at an altitude of almost 2,240 metres, Mexico City is quite dry, and two horses died of exhaustion brought on by severe loss of electrolytes due to the intense heat and the demands of the cross-country course.

Despite its less than ideal subtropical climate, Hong Kong possesses not only the necessary certification as an equine

Fig. 3.5 An accident takes place in a Yunnan minority horse race in April 2013.

disease-free zone but, most importantly, superlative and cutting-edge facilities which include an equine hospital and an equine drug-testing laboratory.

Learning from experience gained at the Olympic Games in Atlanta, Barcelona and Seoul, where stressful environments played a role, the Hong Kong Jockey Club, the organizer in Hong Kong, took additional precautions to assure the safety of equine athletes.

Investing over HK$109 million (US$14m), four blocks of air-conditioned stalls capable of housing 200 horses and an indoor air-conditioned training venue were newly constructed. The air-conditioned stalls were also equipped with fans that kept most of the outside hot air near the entrance and from circulating with the inside air. These facilities allowed the horses to acclimatize over a short period and to perform at their optimal levels.

Cooling facilities were seen at every corner of the venue. Enlarged golf carts equipped with portable generators alongside hoses and bathtub-sized plastic tanks full of cold water were ready to drench overheated horses. Rows of powerful fans fitted with water valves were set up to regulate the horses' temperature before and after the competitions.

In that sense, the riders and the horses competed in a safe and caring environment.

RISK FACTORS IN RIDING

While sports can be made safer by taking all the necessary precautionary measures, there remain however hazards inherent in equestrian pursuits, which are recognized as being among the most dangerous sports in the world.

In China, sports-related horse accidents are, for the most part, caused by insufficient safety measures as equestrian sports are not well regulated. This is especially the case in rural areas where helmets are rarely worn by riders (Fig. 3.5).

In most sports-related accident studies, equestrian pursuits scored the highest mortality and hospital admission rates. Sadly, many horses are thus killed and injured in China, as elsewhere in the world, and advocacy groups have been calling for a ban on certain sports. In ancient China, many literati attempted to get polo banned as the sport itself was perilous. Equestrian sports are considered of higher risk than football, skiing and basketball, and are rated alongside motorsports and powerboat racing.

What is deceptive is that the activities fare under a rather safe façade. Unlike motorsports, with their obviously dangerous image – first aid attendants always standing by – of which participants are highly conscious, horse riders always dress up, appearing poised and elegantly composed. Even the horses look docile – most often!

The truth is a different story. Mounting a horse elevates the rider's head at least two metres from the ground on an animal that weighs 500 kg or more, can kick with a force of nearly 1,000 kg and achieve running speeds of up to 80 kilometres an hour. Horses are less predictable and more sensitive than either a motorcycle or racing car. They have such a highly developed "fright and flight" instinct, often whinnying at the slightest loud noise, or freaking out about something unexpected, like an umbrella. Their fears will surface unpredictably and explosively to trigger an accident.

Despite its danger, the sport nevertheless is still appealing. Human nature provides us with the desire to take risks, exhibit aggression and seek thrills. It is simply a demonstration of the survival instinct which allowed our ancestors to ward off predators and explore new territories. It is part of our life. What better way to fulfil one's lust for competitiveness with the element of peril? It is a sport where mistakes can be as costly as the loss of life, a leg or an arm.

And that's not all. In the stressful moments, an adrenaline rush is induced, empowering us with extra oxygen, energy and euphoric hormones to confront dangers. It is intense, pleasurable and above all addictive! Some people even claim that there is no other feeling in the world that is as satisfying. It is the beauty and mystique of dangerous sports.

Fun, skill and thrill are all parts of the game. They make equestrian pursuits exhilarating, colourful and fulfilling.

CHARIOTEERING

Sports in ancient Chinese society served a number of purposes – military, social, educational and health. In the Western Zhou dynasty (1046–771 B.C.), charioteering, rites, music, archery, calligraphy and mathematics were considered an important constituent of civil and military education for young aristocrats. According to Confucius, men who excelled in these six arts were deemed to have reached the state of perfection as gentlemen.

Fig. 3.6 Stone carving from the Han dynasty depicts a speeding carriage.

CHARIOT DRIVING AS AN ART

Chariot drivers were required to drive with skill. The primitive design of the chariot and the rugged terrain in ancient China demanded accomplished dexterity. On top of this, emphasis was placed on the cultivation of virtue, self-improvement and recreational pleasure. The *Rites of the Zhou*, a classic detailing the government system of the Zhou dynasty, defines the five important chariot driving skills in the following way.

Ming He Luan: When a carriage is moving, the sound of the bells hanging on the front and side poles of the carriage should be light and rhythmic, demonstrating a steady speed.

Zhu Shui Qu: Drivers should manage to pass through complex and dangerous terrain, such as negotiating meandering river banks at high speed without falling into the river.

Guo Jun Biao: High speed driving encourages carelessness and a neglect of the basic rites and morals. Drivers are thus required to learn those rites which are applied in front of emperors so that they can drive steadily and safely, demonstrating calmness and virtue.

Wu Jiao Qu: Drivers should not swerve recklessly while driving through crossroads busy with pedestrians and carriages.

Zhu Qin Zuo: This is the highest level of driving skill. When out hunting or on the battlefield, drivers should force prey or the enemy to the left so as to create a favourable position from which archers can shoot.

To achieve these standards, intelligence, physical strength, courage and most of all virtue were all needed.

Charioteering became popular from the Warring States period (475–221 B.C.) to the Han dynasty (206 B.C.–A.D. 220). Scenes of chariots being driven at high speed are depicted in various archaeological relics, such as murals, painted stone and brick friezes (Fig. 3.6) reflecting the ongoing development of the sport. The *Shi Jing* (Book of Songs) suggests that chariot races were held among aristocrats who liked to wager on the sport.

With the more widespread use of horses and the improvement in manège, the importance of charioteering as a physical activity diminished.

RACING

The earliest depiction of horse racing was discovered in a rock carving in Inner Mongolia, dating from the Warring States period.

Fig. 3.7 Young jockeys take part in the Naadam Festival.

"Tian Ji Horse Racing" is a famous ancient Chinese story about equine racing that happened during that period. Horse racing and the betting that went with it were a very popular entertainment among aristocrats in the Qi Kingdom, and General Tian Ji unfortunately lost a lot on the racecourse. Sun Bin (died 316 B.C.) was one of the greatest military strategists in history and wrote his own military treatise, the *Sun Bin Military Treatise*. He was badly disabled through political persecution in the Chen Kingdom and fled to the Qi Kingdom, first becoming an adviser to General Tian Ji.

Race horses in the Qi were divided into three classes according to performance and each class had its own special harness. Sun Bin found that the quality of General Tian's horses was not inferior to the more successful horses but that his failure was simply due to his inappropriate racing strategy. Sun Bin was so confident of his winning that General Tian invited King Qi to attend a high-stakes race.

On the day, Sun Bin had General Tian's third-class horse equipped with first-class gear. As expected, General Tian lost in the first leg. He then had his first-class horse fitted with second-class gear and his second-class horse with third-class gear. Not altogether surprisingly, General Tian won the last two legs.

Sun Bin had actually cheated. But General Tian had finally won!

Sun Bin's strategic military brilliance was highly appreciated by King Qi and General Tian. However, his maimed legs made him ineligible for the role of head commander as it was a rule that all commanders must be horseback riders. Between them, the King and General Tian decided to appoint Sun Bin head advisor to the army instead.

The Yuan

The rulers of the Yuan dynasty (1279–1368) were Mongolian nomads and it goes without saying that horse racing was very popular with them and among their people.

The Naadam Festival has been one of the most important dates in the Mongolian national calendar for centuries. At the time of Genghis Khan in the 13th century, it was a religious event when the appointment of new officers, resignation of others, administration of rewards and punishments, as well as the triathlon of wrestling, archery and horse racing all took place. The triathlon gradually became the focus of the festival. Traditionally, the event is celebrated on the fourth day of the sixth lunar month (between July and August in the solar calendar). Nowadays, the biggest Naadam Festival is held every year in the Outer Mongolian capital Ulaanbaatar in celebration of the National Holiday from July 11th to 13th. Other places across Outer Mongolia and Mongolian-inhabited parts of China, like Inner Mongolia, Beijing and Harbin, organize their own, smaller Naadam festivities.

In Naadam, horses are divided into six categories by age starting from two-year-olds. Races in Western countries generally cover no more than three kilometres, whereas the race in Naadam is a cross-country event covering some 15 to 30 kilometres depending on the class. For example, two-year-old horses compete for 16 kilometres and seven-year-olds for 30 kilometres. The race is primarily designed to test the skills of the horses: the riders are children aged five to 13 years, some riding bareback (Fig. 3.7).

Having said that, the riders are still trained for months and the horses are fed special diets before the race. The champion horse earns the title of "leader of ten thousand horses", or "champion over a thousand wings", among others. Along with the other top four winners, they are praised in poetry and music and splashed with fresh milk or yoghurt. Cultivation of virtue and self-improvement coupled with skills are emphasized, so that both winners and losers gain respect. In the race for the two-year-old horses, the losers are still serenaded with a song in order to encourage them to do better next time.

The Qing

Founded by the nomadic Manchu, the Qing dynasty (1644–1911) also recognized the value of horsemanship. During the massive annual hunting event Mulanquimi, horse racing and lassoing contests were held for child riders aged six or seven. Given that the emperors presented

Fig. 3.8 Tibetan horse racing festival. A champion horse and its rider are draped in traditional *khatas* (ceremonial scarves), which symbolize purity and compassion.

The first race meeting was probably staged around 1798–1799 in Areia Preta, Macau, which was already a de facto colony of the Portuguese Empire but remained under Chinese authority and sovereignty.

The East India Company, the world's most powerful company at that time, had based itself in the trading port since 1773. This great opium trader maintained the Bengal Army with its cavalry and had obtained the finest horses from across Asia. Britons working for the company and those engaged in the China–India trade in Macau made up the participants, be they owners, grooms, trainers, riders, bet placers and audiences of the races. In a nutshell, the race was very much an East India Company event.

Horses were imported from Calcutta, and later, there were ponies from Manila in the Philippines and Arabs, including very fine ones from India.

After the Treaty of Nanking was signed in 1842, Hong Kong became a British possession and five South China ports – Canton, Amoy, Foochow (now known as Guangzhou, Xiamen and Fuzhou, respectively), Ningbo and Shanghai – were opened for foreign trade. Under other unequal treaties, Tientsin, Peking (now Tianjin and Beijing), Wuhan, Jiujiang, Zhenjiang, etc., also became concession territories which were governed and occupied by foreign powers. These concessions followed in Macau's footsteps to stage race meetings even during the wars.

In 1844, when a dispute between the Portuguese and Hong Kong governments over Macau's sovereignty broke out, Britons were no longer allowed to enter the Portuguese possession. As the key participants disappeared, no more race meetings were held there.

Hong Kong staged its first race meeting in 1845. Its Happy Valley racecourse was initiated by Governor Sir John Francis Davis and the Jardine family which owned the licence for the course.

the prizes themselves, the competitive atmosphere was very intense and contributed to enhancing the standard of equestrianism.

Horse racing is also part of the heritage of many other ethnic minorities in China. Tibet has seen annual horse racing since the 15th century, rewarding the champions with prize money (Fig. 3.8). The Bai, Yi and Kazak ethnic minorities have also held horse racing events for centuries.

Modern Horse Racing

In most countries, modern horse racing is seen as a flagrant display of prosperity or just a hobby. In China, it means much more.

Fig. 3.9 Notable in this scene of a jockey on a Sanhe horse at a Shanghai racecourse in the early part of the 20th century is that the spectators' stand contains only Western racegoers.

Likewise, racing in Hong Kong was pretty much an event of Jardine, Matheson and Co., which was also a leading opium trading house. Among all the concessions, Hong Kong was considered second only to Shanghai in terms of race wagers and prize money.

Shanghai commenced its first race meeting in 1851. By 1864, it already had four racecourses. Horse traders back in Mongolia all knew the Shanghainese were very generous. If a Shanghainese paid 25 taels for a Sanhe horse, other Chinese would pay 20 taels at the most (Fig. 3.9).

Strictly speaking, all of these were only amateur races, but of exceptional horsemanship. Owners or sons of owners, who were mostly aristocrats, could also be riders. David Sassoon, a Baghdadi Jewish opium trader and an owner of the Leviathan Stable in Shanghai, was one of the outstanding riders in both Hong Kong and Shanghai of his generation. The Leviathan Stable dominated the Shanghai races from 1886 to 1894 and lifted the racing standards.

John Johnston was another successful owner-rider of the early 20th century from the Jardine family. His wife was one of the first lady owners admitted to the Hong Kong Jockey Club in 1921. John Johnston rode her Irish Stew to win the Valley Stakes in Hong Kong, while he also won numerous victories back in Shanghai.

The third Hong Kong governor, Sir George Bonham, had his own stable. He raced on his own horses and also others.

Likewise, there were many Chinese owner-riders. C.T. Chu was the first Chinese star owner-rider and scion of Chu Pao-san, a Ningbo millionaire who actively promoted the establishment of the International Recreation Club, the first racing club jointly formed by Chinese and Westerners.

Riders and horses would travel by river steamers to race in other concessions. In the late 19th century, a return journey for a Shanghai horse to Hankow was 1,900 kilometres by river, while a return journey for a Hong Kong horse to Shanghai was 3,400 kilometres by sea. Travelling such a long distance to compete sounds incredible by those days' standards.

Before 1937, racecourses were primarily established by Westerners, Japanese or were joint ventures between Chinese and foreigners; all operated to a very high standard. Later, more and more Chinese began to participate, resulting in the emergence of Chinese-operated racecourses, just like the International Recreation Club in Shanghai and the Hankow International Racecourse in Wuhan.

Following its invasion of Manchuria, the Japanese army occupied north-east China from 1932 to 1945, setting up the Manchukuo kingdom. During that period, 10 racecourses were built with horses supplied from 10 studs which bred not only hot bloods for racing, but also warm bloods and cold bloods for military purposes. Equine veterinary training was also put in place. All the profits from horse racing ended up in financing the puppet government.

COMMUNISM AND HORSE RACING

Horse racing completely disappeared in 1949 when the Communist Party came to power. Along with all kinds of gambling, prostitution and drug dealing, horse racing was banned as it was considered decadent and a throwback to colonial days. It was not until the early 1990s, when national races were organized and jockey clubs were established that the sport reappeared on the mainland as a race-only spectacle. By then, Nanjing, Wuhan, Beijing, Guangzhou, Dongguan, Shenzhen, Jinan and Ningbo had each built a racecourse.

At present, only Hong Kong and Macau in China have government approval to hold horse racing where gambling is allowed. The sport reflects the heritage of colonization in different periods and is likewise a challenge to the political ideology of the country.

WUHAN'S HISTORY

Since 2003, Wuhan has held the country's biggest equestrian festival, which takes place annually in October. The city is the capital of Hubei, a province which lies in the middle reaches of the Yangtze River. The modern metropolis is the largest city in central China and is recognized as the political, economic, financial, cultural, educational and transportation hub of the region.

By the early 1900s, Wuhan already had a total of three racecourses. The first Hankow race meeting was held in 1864 in the Hankow Race Club and Recreation Ground (Hankow Racecourse), which was operated by a British businessman. Unfortunately, Chinese residents were badly discriminated against and only allowed to watch the races from a remote area. Signposts saying "Chinese are not allowed to enter" were installed within clear zones.

Liu Xinsheng was the landlord of the racecourse and was so angered by the discrimination that in 1906 he aligned with a banker, Liang Junhua, and 34 other patriotic partners to build a much bigger racecourse, calling it Hankow Chinese Race Club & Recreation Ground (Hankow Chinese Racecourse). Unlike the original Hankow Racecourse, Hankow Chinese Racecourse did not have an entry fee or impose any discriminatory measures.

In 1924, Wang Zhifu, Wu Chunsheng and other partners established the Hankow International Racecourse, which became the second largest in Wuhan. However, Hankow Racecourse still managed to outperform its two rivals with daily wagers in its heyday of up to 200,000 silver dollars (national currency in the 1920s and 1930s).

In the early 2000s, Wuhan's Orient Lucky City was among the biggest racecourses in mainland China. The site spreads over 100 hectares, with an oval-shaped dirt track of 1,620 metres in length and 28 metres in width, a stable capacity for 2,000 horses, a stadium capacity for 30,000 spectators and a 480-metre video screen.

During the 2008 equestrian festival, a lottery was introduced into the competition and was seen as the first official horse bet in mainland China since 1949. Technically speaking though, it was a horse-racing lottery as fixed-odds betting was not involved. The authorities bent over backwards to emphasize that it was a competition of intelligence along with the lottery so as to play down any accusation of gambling, which the Communist Party has long prohibited on moral grounds. Punters were only allowed to place free bets and instead of cash prizes, those who won were given 20 instant scratch-off tickets by the local sports lottery administration. One year later, the prizes were prepaid cellphone recharge cards. Before June 2011, some five to seven internal trial races were run twice a week, which were not open to the public. Since then, open races have been staged. In 2018 and 2019, 20 open meetings, each comprising five races, were held each year in Wuhan on certain Saturday afternoons. Spectators were allowed to participate during and between races by playing interactive games on mobile devices to win prizes.

RACING ELSEWHERE IN CHINA

Horse racing also took place in other cities. However, by 2000, with the exception of Jinan, Nanjing and Wuhan, all had been shut down by the government under the "Strict Forbidden Order on Horseracing Gambling" put forward jointly by the Ministry of Public Security, the Ministry of Supervision, the State Administration of Industry and Commerce, the General Administration of Sport and the National Tourism Administration.

Added to the mix are notices on "Enforcement for the Catalogue of Restriction of Land Use Items" and "Enforcement for the Catalogue of Prohibition of Land Use Items". In December 2006, the Ministry of Land and Resources and State Development and the Reform Commission jointly released these notices to outlaw the use of land for the construction of racecourses.

Nanjing and Jinan racecourses were built for the 10th and 11th National Games staged in 2005 and 2009, respectively.

The National Games, which was inaugurated 50 years ago, is a top national level sports spectacle held every four years in different cities. It brings together prime athletes from across the country to compete against one another in a variety of individual and team sports. The equestrian competition is one of the premier events of the National Games.

However, after the National Games, the racecourse in Nanjing did not stage any more major events and has since been turned into a car park. Jinan racecourse is operated alongside other tourist attractions without any meetings.

The racecourse at the Guangzhou Racing Club was built in 1993, about a year after Deng Xiaoping's southern visit in 1992, and was formerly of some significance. The venue spread over 33 hectares and featured a nine-storey building, a three-level grandstand, a 1,695-metre-long and 32-metre-wide turf track, a 1,485-metre-long and 25-metre-wide dirt track, robotic cameras, finish-line timers, a huge video screen and other cutting-edge facilities.

It offered the first horse lottery approved by the government since the Communist Party's ascent to power in 1949. The club was a non-profit organization which donated all its surplus funds from racing to charity and community activities. Three meetings were held every week and the racecourse attracted record one-day crowds of about 20,000 with wagers going as high as RMB12 million (US$1.7m). During its existence the club donated some RMB300 million (US$44m) to charity and provided thousands of job opportunities.

However, after years of operation, gambling inevitably came about, seriously violating government policies against decadence. In 1999, the authorities started investigations into the club's management and found that it had in fact incurred an accumulated loss of RMB600 million (US$88m) in six years. Far worse, the chairman, Huang Qihuan, was determined to have been involved in bribery, embezzlement and fraud, and ended up with a 19-year custodial sentence. Racing activities were forced to stop by the end of 1999 and Guangzhou horse racing was labelled as an unsatisfactory experiment because the authorities had been unable to prevent people from placing bets. This was such a blow to the fledgling horse racing business in China.

The racecourse is still there, but is now an automobile, catering and entertainment centre, with over 100 shops. The club still owes RMB1.2 billion (US$176m) to its creditors.

The central government's clampdown, nevertheless, has not suppressed the enthusiasm of investors and local governments who live in hope of horse racing being liberated one day. Since 2009, a few new racecourse developments have been embarked upon; among these are two grand projects, bolstered by global heavyweights, in Sichuan and Tianjin.

In Chengdu, the capital of Sichuan, a racecourse reportedly of international standard and jointly built by local authorities and investors, has since 2014 annually hosted the "Chengdu Dubai International Cup". Under the clout of Sheikh Mohammed, the Cup managed to secure sponsors, feature world-renowned jockeys and distribute lucrative prize money. The question is whether this is sustainable going forward without any betting.

Tianjin Horse City is an ambitious US$2.6 billion project, backed by Coolmore and Teo Ah King's China Horse Club, reportedly to contain an international equestrian college, a breeding base, an auction base, a feedstuff plant, a racecourse and a five-to-seven-star hotel with a formidable phoenix-shaped grandstand.

Seeing it as carrying a strong economic tailwind, the local government is eager to embrace the pursuit despite the red light from the central government and the immaturity of equestrianism. The project was announced in 2010, but there was no update as at 2020.

Separately, as an exemplary organization in benefiting society using the proceeds of horse betting at a world-class standard, the Hong Kong Jockey Club made its move into China by building a training base in Conghua, Guangdong. The world-class racecourse spreads over 150 hectares and can accommodate more than 660 horses. It is a training centre for Hong Kong race horses as the Club succeeded in creating an internationally recognized equine disease-free zone stretching from Hong Kong to Conghua. In March 2019, the racecourse staged its first exhibition meeting, at which there was no betting. The Club expects to hold another meeting in 2020, while insisting it does not have any thought to betting in China.

LIBERALIZATION OF HORSE RACING

In Chinese society, it is almost impossible for horse racing, unlike football and basketball, to thrive as a sport if it does not involve gambling.

Liberalization of horse racing is not without its upside. It could create huge numbers of jobs once a nationwide betting network is fully rolled out and generate a whopping amount of annual revenue.

As of 2019, revenue from legal lotteries was worth some RMB422 billion (US$60bn), of which RMB191 billion (US$27bn) was attributable to the

Fig. 3.10 Mural of a polo match. 1.65 x 6.75 m. Shaanxi History Museum, Xi'an. This fresco in the tomb of Crown Prince Zhanghuai depicts more than 20 Tang imperial family members playing polo.

welfare lottery and RMB231 billion (US$33bn) to the sports lottery. However, this figure is dwarfed when compared with that of illegal gambling. The estimated wagers from illegal gambling, which includes internet betting, underground casinos and private lotteries, are multiple times the amount the legal lottery takes. Not only is illegal gambling on its territory a huge headache for the Chinese government, the gambling away of public money, especially by senior Chinese officials and heads of state-owned enterprises, is rampant. Quite apart from the capital outflow to Hong Kong and other nearby countries like Australia, South Korea, Myanmar, Cambodia and the Philippines for gambling, Chinese visitors accounted for approximately 70 per cent of the gross revenue of the Macau casino industry, which reached its height at 361 billion patacas (US$45bn) in 2013.

Although the Chinese government started to crack down on corruption and launched an anti-extravagancy campaign in 2014, the 2019 gross revenue at Macau casinos still came in at 292.4 billion patacas (US$37bn). The staggering capital outflow, which was partially sustained by money laundering and racketeering, brought scant or no economic return for the nation.

It is argued that placing a wager on the outcome of an event is a part of human instinct and is unstoppable; and by liberalizing gambling, fewer people will turn to underground or extraterritorial markets. Enormous revenue from deregulated gambling could be used to fund a wide range of infrastructure and social welfare programmes.

China remains the world's largest untapped gambling market, yet the government is reluctant to relent on betting and officials assert that gambling will never be the country's economic backbone.

As a communist country, ideological barriers are substantial to the development of the gambling industry, which is seen by communist doctrine to be the root and branch of capitalist instability and injustice. In particular, gambling is often associated with corruption, embezzlement and money laundering, while it is a well-known fact that the Chinese are fond of gambling and that Chinese gamblers often find it difficult to restrain their habit.

Given that China lacks a sophisticated regulatory framework to ensure secure and transparent gambling and to prevent gambling-related crimes from taking place, these issues potentially form and drive a vicious circle, ending up with social unrest. Anti-graft has been a pre-eminent cause of the Chinese government since President Xi Jinping took office in 2012. Under the current political agenda, horse racing on the back of all these critical concerns is unlikely to be liberalized.

Adding to the list of unfavourable conditions is China's lack of animal welfare legislations to protect race horses from abuse and cruelty. Many horse-related standards, facilities and professionals which could help ensure the welfare and rights of race horses, all still have more than enough room for improvement. These shortcomings range from manège, stable management, quarantine, husbandry, transportation, medicine and insurance to vets, grooms, farriers, breeders and trainers.

Importing to or breeding a great number of race horses in China without enough protection for them will just create huge issues for the nation. Until an international standard of horse racing can be attained, there will be fierce criticism from international and domestic advocates of animal welfare. This will simply create a lot of hassle and none of the expected monetary benefits.

In contrast, basketball and football are already very popular in China, even Xi Jinping is a football fan. International-standard stadiums together with supplementary and complementary amenities and facilities of the two disciplines have also been established. Most of all, these sports are not embroiled in animal rights issues. If China seeks to take advantage of sports betting, in all likelihood, basketball and football stand a higher chance of getting the green light.

POLO

No one really knows where or when polo originated. However, the sport's modern name can be traced to Tibet, where it was known as "Pulu", meaning either ball or willow root, from which the polo balls were crafted in the country over 2,000 years ago.

Polo was called *jiju* (striking the ball) in ancient China. The earliest account in China appeared in the *Mingdu Pian* (Song of the Renowned Capital), by Cao Zhi in the early Three Kingdoms period. In the poem, Cao Zhi described young people from Luoyang, capital of the East Han dynasty, in flattering outfits playing polo until sunset and wrote also of the polo pitches and the prowess the game entailed.

Polo requires horsemanship, courage, hand-to-eye coordination and team spirit from both horses and players alike. These are all skills akin to those needed on the battlefield, which is why polo was often deployed as a major military training exercise.

CHINA'S GOLDEN AGE OF POLO

The golden age of polo in China occurred in the Tang dynasty, the founders of which already had a profound equestrian heritage and all 16 emperors and aristocrats were polo aficionados – and some of them were even prodigies (Fig. 3.10). The popularity of polo can also be attributed to the strength of the cavalry forces and the extensive cultural exchange.

Polo emerged as a link between equestrianism, exercise and the military. The Tang military employed it as a form of training, much as the Han had played football. All provincial governors and military bases had standard polo pitches for training and leisure purposes, and polo matches were part of troop reviews. Officials with exceptional skills and flair would often be promoted, such that the sport provided ordinary people with a form of social mobility.

At that time, polo was a national sport in Tibet with a sophisticated standard that was well known among neighbouring countries. Accompanying the closer diplomatic relationship between the two royal houses, the game also flourished in the Tang. When two Tang princesses were married into Tibet, polo balls of gold were among the wedding gifts from the Tibetan court. Tournaments between nations were not uncommon. In a tournament against Tibet, the Tang first fell behind considerably until Tang Xuanzong, who was then a prince, was sent in as a replacement and finally saved the day.

Many aristocrats had their own private pitches, while intellectuals played polo matches as one of

Fig. 3.11 The Tianjin polo field bordered by luxurious condominiums.

the celebratory events for their success in imperial examinations. Studs were built to breed the best polo horses, poems were written about the sport, and plenty of artefacts like bronze mirrors, murals and figurines were created, testifying to the popularity of the spectacle.

Polo pitches formed part of palace development plans and were equipped with top-notch facilities. Polo fields, measuring 1,000 paces by 100 paces, were dedicated level surfaces, often treated with oil to prevent slipping in rain and the dust from flying about. They were surrounded by three short walls and a lookout pavilion, stage, hall, or tower.

Polo fields in military bases were huge. According to historical records, the Hunan military base provided hospitality to 3,000 armies from Xuzhou in a polo field, and the Anhui military base had hosted a banquet for 1,000 armies in another.

Ancient China held a liberal attitude towards women's participation in sports, particularly in royal circles where aristocrats allowed their maids to exercise in the backyards. Pursuits played by women, such as polo, were in vogue.

Records also show that during the reign of Emperor Taizong Li Shimin, polo teams were formed with 50 maids as members and eunuchs as coaches. Wu Zetian (624–705), who was later the fifth monarch in the dynasty and the sole empress in Chinese history, was made a team captain. In a polo match staged as a performance for the emperor, the teams both wore men's clothing and white boots: one was dressed in red satin with red flowers in their hair, and the other in green satin with green flowers. To accompanying instrumental music, the teams rode into the court, making three circuits and paying homage to the spectators. They finally stopped in front of the emperor, dismounted and shouted three times: "Long live the emperor". After these rituals were completed, the match started and proceeded in the best of spirits on both sides with vociferous support from spectators. The thrilling match finished with the red team led by Wu Zetian beating their opponents 2-1.

A variant form of polo was *luju*. It was often played by children and women on donkeys, which were smaller and considered less violent and more manageable than horses. Luju was popular during the Tang and Song dynasties.

Playing polo was seen as very dangerous and many literati believed that it should be forbidden. However, its popularity among emperors and its strong link with military preparation meant the plea was made in vain: its prevalence continued.

It was at the time of the Southern Song dynasty (1127–1279) that the sport was gradually withdrawn from the field of military training, but it remained popular as a court entertainment and recreation.

Although the Yuan dynasty was founded by horse-riding nomadic pastoral tribes, polo was not part of their heritage. But with their strong equestrian background, they fell in love with the sport soon after their successful overthrow of the crumbling Song dynasty.

QING TO THE PRESENT

During the Qing dynasty, private martial art practices and horse breeding were proscribed. Although prohibition of horse breeding was slightly relaxed 10 years after the precocious boy emperor, Kangxi, was enthroned (1671), polo playing among the general public continued to be prohibited. Matches played by girls occasionally however came about in social gatherings. In the final years of the Qing rule when foreign forces arrived in China and gunboat diplomacy prevailed, the cavalry was phased out, overcome by the superior weaponry and technology of the Western powers. Polo playing by the populace, the court and the military declined in tandem with the imperial era.

In the 21st century, China's booming economy is creating upwardly mobile "new aristocrats" with wealth and talent to flaunt. History has inevitably repeated itself, but in a different manner: smart polo clubs have been set up, while international events with sponsorships from luxury brands and property developers are held amid all the trappings of pageantry. Housing complexes have been built alongside polo fields, displaying panache and, most of all, lifting property profiles (Fig. 3.11). The strong link between polo and military preparation is now replaced by wealth, status and business opportunities.

Despite this, polo is still a very exclusive hobby for a small number of people who are rich, horsey and sporty. In fact, because the numbers of polo participants, spectators and lovers of the game are so limited, the majority of Chinese nationals have little idea about the sport. As a gimmick for boosting the attraction of residential property, the hobby has turned out to be less appealing than expected.

HORSE DANCING

Horse dancing in ancient China was different from modern day dressage. Horses danced to music individually or in groups without riders.

Dancing horses were mainly tributes to the imperial courts from western regions, such as Fergana and Tuyuhun (present-day Qinghai), but also from places much further afield such as Rome. They were of good quality and well trained for dancing before being imported to China.

Horse dancing can be traced back to Emperor Han Wudi (reigned 140–87 B.C.) who received a Fergana horse trained to bow and move to the rhythm of drums.

During the Three Kingdoms (220–280), Cao Zhi gave his elder brother, Emperor Cao Pi, a Fergana horse which could dance to drum music.

The earliest dancing troupe dates back to 458 during the Southern and Northern dynasties when a herd of performing horses was bought to the Liu Song court as tribute.

Although it originated elsewhere, the art of horse dancing culminated in China and reached its incomparable zenith in the Tang dynasty (Fig. 3.12). This was reflected by the establishment of a specialized organization to keep and train the horses, take care of their diet and harnesses, and develop sophisticated choreography and music for their performances.

In 729, Tang Xuanzong declared his birthday, the fifth day of the eighth lunar month, as the "Thousand-Autumns Holiday", which implied best wishes for a long life to himself.

Programmes to mark the occasion's festivity included military drills, dances and various gala performances. Elephants and rhinoceroses were brought in to enchant and wow the crowds. The highlight of the pomp though was the dance performed by 100 meticulously trained horses whose feats were the most extravagant and spectacular ever seen.

All the prancing horses were richly caparisoned, their coverings patterned with embroidery and fringed with gold bells, their manes and forelocks decorated with pearls and jades.

The horses danced in rhythmic paces, moving their heads, tails and bodies in response to the music's tempo and modulation.

The *Song of the Upturned Cup* was the most acclaimed music for horse dancing. It first appeared in the North Zhou in the Southern and Northern dynasties. A horse dance was choreographed to the festal song, culminating in a final superb scene where the horses drop their heads, flicker their tails, bend their knees and clench wine cups in their mouths to toast the emperor (Fig. 2.15).

The embellished horses danced not only on the ground, but also on platforms of several tiers. They were driven to the top of three-tiered wood-plank platforms where they turned and twirled around as if flying; at other times, sturdy men were ordered to lift a bench for the horses to dance atop.

The Thousand-Autumns celebration continued for another 23 years. The rebellion of An Lushan, which occurred in 755–763, seriously weakened the dynasty and resulted in the decline of the exclusive court entertainment. During the political upheaval, some horses were acquired by Tian Chengsi, a general under An Lushan. He had never seen horse dancing and was not aware of the animals' special talents. He kept the dancing horses together with ordinary horses. One day, when the soldiers of his army were enjoying a sacrificial feast and playing music, the dancing horses, driven by a reflexive response, began to dance. The servants and lackeys considered them bewitched and beat them with stable brooms. The horses did not stop and sadly they were flogged to death.

The dancing horses vividly symbolized the flourish of Xuanzong's court and the glamour of the Tang equestrian culture. Although their grandeur will never again appear at a Chinese imperial court,

Fig. 3.12 Ceramic dancing horse and its trainer.

they remain the unrivalled performers of the equestrian stage.

THE MODERN EQUESTRIAN DISCIPLINE

Ancient equestrian pursuits had their lows and highs throughout the imperial period. Today, to complement the nation's current rising global economic and political clout, China has invested heavily in selecting and training its athletes in a wide range of disciplines. In the process, it has become a sports superpower, as proved by its dominance at the Beijing Olympics in 2008. Alas, the modern equestrian discipline was not included in its medals.

At present, equestrianism in China is not sufficiently sophisticated to compete on the world's stage. In particular, the necessary soft infrastructure is yet to be put in place.

As the second-largest economy in the world, there is no doubt that China and Chinese riders can afford to develop world-class hard infrastructure, which includes racecourses, arenas, harnesses, riding gear, stables, training facilities and much more.

Soft infrastructure however cannot be secured by money alone: considerable time and effort alike are needed to nurture it. As the major participants of equestrian pursuits, the sport horses are not disposable commodities but sentient athletes which need their riding partners and owners to truly care for their welfare. In general however the lack of success in equestrian sports in China, as elsewhere where much more investment is put into horse racing than other competitive equestrian events, does come down to money.

The four major problems with the modern equestrian disciplines are the over-breeding; mistreatment during training, performances and the off-season; lack of veterinary care; and the way in which unwanted horses are culled.

To address these issues, it is essential to enact animal protection laws, adopt international standards of equine health care and training, acquire proper husbandry knowledge and practices, and most of all to cultivate a sense of respect, compassion, empathy, justice and consideration for the animals among the general public.

The benefits for riders also matter. They should be provided with appropriate training and covered by insurance. Safety laws with such clauses as interdiction of drink riding and the mandatory wearing of helmets should be enforced.

There is still a long way for the Chinese equestrian team to go before being a force to be reckoned with in the global arena. Nevertheless, once all the right soft and hard infrastructure is in place, China will be seen in the eyes of the world and among animal rights activists as having championed a magnificent cause that dates back to the dawn of its history. Medals will be of secondary importance.

CHAPTER FOUR

HORSE BREEDS IN CHINA

INTRODUCTION

In 2020, China reported 47 horse breeds to the Domestic Animal Diversity Information System of the Food and Agriculture Organization of the United Nations. We can classify the breeds into three categories as follows:

Native Breeds		Developed Breeds		Exotic Breeds	
1	Barkol	24	Bohai	41	Akhal-Teke
2	Bose	25	Guanzhong	42	Ardennes
3	Chaidamu	26	Heihe	43	Don
4	Chakouyi	27	Heilongjiang	44	Kabardin
5	Datong	28	Henan Light Draught	45	Orlov Trotter
6	Elunchun	29	Jilin	46	Soviet Heavy Draught
7	Ganzi	30	Jinzhou	47	Thoroughbred
8	Guizhou	31	Kerqin		
9	Hequ	32	New Lijiang		
10	Jianchang	33	Sanhe		
11	Jinjiang	34	Shandan		
12	Kazakh	35	Tieling Draught		
13	Lichuan	36	Xiangfen		
14	Mongolian	37	Xilinguole		
15	Ningqiang	38	Yili*		
16	Tibetan	39	Yiwu		
17	Wenshan	40	Zhangbei		
18	Xinihe				
19	Yanqi				
20	Yongning				
21	Yunnan				
22	Yushu				
23	Zhongdian				

*Misspelt as Iyi in the DAD-IS database, confirmed by the National Animal Station, Ministry of Agriculture, People's Republic of China (PRC).

Of these, 23 are native and include the Mongolian and Hequ horses, and the Southwest ponies (such as the Bose, Guizhou, Jianchang, Lichuan, Ningqiang and Yunnan), 17 are developed breeds and include the Sanhe and Yili horses, and seven are exotic, like the Akal-Tekes and Thoroughbreds. Native and developed breeds account for 90 per cent of the horse population.

In terms of height, most of the horses in China fall into the category of "pony" rather than "horse" based on the Western standard. The Mongolian horses, Hequ horses and Yili horses all measure less than 14.2 hands (144 cm). Sharing a similarity with ponies, they are noticeably tougher, stockier and hardier than large horses, meaning that less food, less care and less time are devoted to their upkeep. They also make ideal trekking companions and mountain "porters" as they are very calm, gentle, alert and intelligent, and above all very sure-footed. Despite this, for various reasons of tradition or other factors such as specific conformation and temperament, they are still classified as horses in China.

Native and developed breeds in China are mainly cold bloods and warm bloods. The classification is not based on the horse's body temperature but on its temperament and the tasks it carries out. Draught horses like South-west ponies are cold bloods and are characterized by their calm and gentle disposition and adaptation to harsh conditions. Although they lack the sheer size and power of typical European cold-blooded horses, such as the Shire and Clydesdale, they still pass as cold bloods. Hot bloods are sprinters and distance runners, with strong wills and fiery tempers. The Thoroughbred and Akhal-Teke, which are exotic breeds, are hot-blooded horses. Horses which do not fall into either of the above two classifications are considered warm bloods. They are calm animals and easily trained, less impulsive than hot bloods while less phlegmatic than cold bloods; they primarily undertake the role of sport horses or light draught horses and include the Yili, Hequ and Sanhe.

China's horses are mainly found in forest grasslands, arid grasslands, mountainous deserts, plain deserts, high-altitude meadows and plateaus in the west, north and south-west. Guangxi, Yunnan, Guizhou, Sichuan, Tibet, Qinghai, Xinjiang and Inner Mongolia husband 75 per cent of the total population. China is a vast country with diverse geographical and climatic conditions which have resulted in a variety of breeds evolving and developing to adapt to the different environmental situations. Distinct regional differences are thus seen among the breeds despite their common ancestry.

From home-grown breeds that are rare or non-existent elsewhere to imported breeds that have influenced native equine bloodlines, they are all part of the Chinese equine history and have a strong bond with the country and the people.

NATIVE BREEDS

THE MONGOLIAN HORSE

The Mongolian horse (Fig. 4.1) is the breed we most often associate with Mongolia, where the people have been famous for their horsemanship since history began. As one of the oldest surviving horse breeds in the world, Mongolian horses feature the largest genetic variety among all the Chinese horse breeds, representing an original nature less affected by human-induced selection. It is also evident that many other breeds descend from Mongolian horses. In China, the horses used by Genghis Khan (1162–1227) are mainly found in Inner Mongolia.

In spite of their diminutive body size, having an on-average withers height of just 12 to 13 hands (122 to 132 cm), they are not ponies as their body build, traditional uses and overall physiology still grant them the title of horses. The breed has a short, muscular neck, compact and stocky body, large head and stout legs. It is found in a wide variety of colours and patterns. The long mane and tail of the animal can be used to produce braiding ropes and violin bows, which have a worldwide reputation.

In China, the sturdy and unpretentious steppe horses are kept by their owners to roam free outdoors, where in Inner Mongolia the summer can reach 25°C and winter fall to -30°C, and are given no hay or other extra feed at any time of the year. The stallions in winter protect their harems and young offspring from wolves.

Primarily employed for riding and carting, the horses are also used for their meat and for the production of *airag*, a fermented horse milk, which is an important staple for the Mongolian people who have few vegetables in their diet.

Fig. 4.1 A herd of Mongolian horses participates in the Equestrian Cultural Festival at Wuzhu Muqin Grassland, Xilingol, Inner Mongolia. The hardy and unpretentious steppe horses are grazed freely in central Inner Mongolia where the summer temperature can reach 25°C and in winter fall to -30°C. The horses are given no hay or other extra feed at any time of the year.

Fig. 4.2 Two snow-white Mongolian horses graze at will by the Genghis Khan Mausoleum in Inner Mongolia. Said to be descendants of Genghis Khan's favourite charger, Yinhebajun, they take part in sacrificial ceremonies, each wearing a Genghis Khan golden saddle.

These small horses are evocative of force, strength and spirit beyond their pony-size physique. A Mongolian horse can carry about one-third of its body weight, while four can haul a load of 2,000 kg for 50 to 60 kilometres a day.

Among the "five treasures" – horses, camels, cattle, sheep and goats – that Mongolian nomads typically possess, the horse is the real jewel of their livestock. Horses and the Mongolians alike grow up with a strong affinity for each other. A Mongolian folk saying has it that if a drunk or injured owner gets onto the back of a Mongolian horse, it will take him home.

On the racecourse, Mongolian horses will try their best – and even die from exhaustion for their owner's glory. Mongolians never shout at their horses or strike them on their heads. They always also carry brushes or sweat-scrapers with them to be able to groom their proudly owned quadruped friends.

During Genghis Khan's extensive conquests, Mongolian horses played a substantial role. The horses' unparalleled vitality and endurance in inhospitable conditions facilitated the long-haul campaigns from Asia to Europe. Their forage requirements were minimal as they subsisted on grass, reducing the baggage burden and military budget of the Mongol cavalry. On top of this, most military mounts were mares, which competed less in the herd and in return caused fewer problems.

Although the breed has created miracles during its history, it does not perform particularly well in modern-day equestrian pursuits. However, it still maintains an exalted position in the hearts of the Mongolian people. Currently, the only living idol in the Genghis Khan Mausoleum is a white

Mongolian horse which is said to be a reincarnation of Wendougenchagan, the representation of the heaven god Sa'erle. When a living idol dies, a foal with a spotless white body, flickering eyes and black hooves, will be chosen as the new reincarnation of Wendougenchagan and will then guard the mausoleum for the length of its life.

Two other much revered snow-white horses graze in the mausoleum grounds and are said to be descendants of Genghis Khan's favourite charger, Yinhebajun (Fig. 4.2). They cannot be ridden or put to work in any way and are left to wander the grounds at will, except during certain sacrificial ceremonies when each takes part tacked up with a Genghis Khan golden saddle.

BIGGEST CHALLENGE

Having in antiquity conquered the world with Genghis Khan, the gallant chargers now face the biggest challenge in their history as a consequence of the environmental and economic upheaval in Inner Mongolia.

In the late 1980s, when desertification and land degradation became acute, every pastoral family was allocated a small piece of land which was demarcated by wire fences. Since then, the horses' traditional roaming areas have been drastically affected: their freedom to meet and mate with horses in other herds, to find safe and sheltered places during natural disasters like blizzards, to choose the grass they want to eat, among other important issues, has been denied them. Mongolian horses have lived free since history began, confining them to small pens is against their instincts. Not only has the mining boom in their traditional grazing lands created huge ecological problems, but the introduction and recent extension of the prohibition on horse grazing challenge the native horse's existence.

Mongolian herders have ended up keeping fewer horses and travelling more by motorized vehicle. Back in 1987, Inner Mongolia had 1.9 million horses. Following grassland degradation and privatization, the number began to decrease drastically. In more recent years, the provincial government and investors have been actively developing breeding, especially for the racing and tourism industries, such that the situation has now improved, with the animal population reaching 638,000 heads in 2018. However, the number of Mongolian horses remains low. The breed currently accounts for only a fraction of the province's total.

Claiming that horses are their life, elderly herders are lamenting their substantially shrinking herds and the deteriorating living environment. Today, Mongolian horses are being developed as a cultural and tourism symbol while desertification reportedly has been slightly stemmed.

THE HEQU HORSE

Hequ means "river bend". The Hequ is a breed developed in the border regions of Qinghai, Sichuan and Gansu through which the S-shaped first bend of the Yellow River runs. The area features both grass and swampy steppes, which yield abundant forage. Before 1954, the breed was known as Nanfan. It is also sometimes referred to as Qiaoke, which means swamp horses in Tibetan. Depending on the specific places to which it is considered native, the breed is also called Ruoergai, Tangke and Kesheng.

The Hequ is distinctive in its type and conformation. The body is well proportioned, the head hare-like, the nostrils wide, the neck long and sloped, the loins of medium length and flat, the back long and the ribs well sprung. Its legs are properly set, though often somewhat coarse, with well-developed joints and strong ligaments, and wide and dense hooves, though these are often cracked. The horse's measurements are approximately as follows: withers height 13 hands (132 cm); oblique body length 139 cm; chest girth 165 cm; cannon bone girth 18 cm; and weight 350 kg. This is a draught and riding animal, able to cover 50 kilometres in a day with a load of 100 to 500 kg, and is especially ideal for use in mountains and swamps.

But what makes the horse really special is its inhospitable homeland.

The Hequ's home is a landlocked area made up mostly of mountains and high plateaus at an altitude of between 3,300 to 4,000 metres (Fig. 4.3). It has a continental climate which is characterized by sharp temperature variations of up to 25°C, both seasonal and daily, and an average yearly temperature of between 0.3 to 1.4°C, with long winters, extremely short summers, frequent winds and most of all, strong ultraviolet radiation.

A product of the habitat's special geography, climate and vegetation, the Hequ are horses of high altitudes, rarefied air, alpine meadows and rolling, hilly areas. They are champions at altitudes higher than 3,000 metres, where other breeds

cannot compete. Most Hequ are black, brown or grey as they have evolved to avoid sunburn in their high altitude habitat. These darker coloured coats offer more protective melanin against ultraviolet radiation and also help to absorb heat (Fig. 4.4).

The Hequ are bigger than most other breeds in China. According to the German biologist Carl Bergmann, the larger body masses contain more cells in which the normal byproduct of metabolism is heat production. As a result, the bigger Hequ can produce more internal heat.

The animals have adapted to the high altitude, low pressure and thin air where they live by developing barrel chests containing larger lungs and they have higher white and red blood-cell counts, the latter providing a heightened oxygen-carrying capacity.

Like many other mountain horses, the Hequ have sickle hocks, which may be considered a conformation fault but actually add a somewhat rolling movement to the animal's hindquarters, helping it to survive in rugged terrains.

The Hequ are very fertile. Mares produce foals every year or two. May and June are the optimum months for mating because when the foals are born a good 11 months later, it is the best time of the year climatically with an abundance of fresh grass rich in protein of the steppe.

THE HEQU IN HISTORY

There is a close resemblance between the Hequ and the terracotta horses found in Xi'an. It is believed that Hequ horses were the primary breed employed by the Qin army and that the terracotta steeds were modelled on them. Historically, the heartland of the Qin is modern day Shaanxi, which is located right next to the western highlands, traditional home of the Hequ.

Fig. 4.3 The Chinese People's Liberation Army (PLA) cavalry patrol on Hequ horses in border areas of the Himalayas, often up to altitudes of 5,218 metres.

The terracotta steeds are either cavalry or chariot horses. The cavalry horses carry saddles and only a proportion of them are geldings, while all the chariot horses are geldings and are not carrying any gear (Fig. 4.5).

Xiang Yu (232–202 B.C.) was a prominent general and leader of the rebel forces which overthrew the Qin dynasty (221–206 B.C.). Legend has it that his horse, Wuzhui, was a Hequ horse. It had a black coat, typical of the breed, with four white hooves. Wuzhui carried Xiang Yu through a succession of battles until he was ultimately defeated by Liu Bang, founder of the Han dynasty. Xiang retreated to the Wu River where a boat was got ready to take him home, but the general was too ashamed to face the people back in his homeland and he refused to board, asking those present to take his beloved Wuzhui instead. The horse however was very reluctant to leave its master and once on board began neighing and leaping about; it kept turning back its head and finally – and sadly – jumped into the river, vanishing without a trace. Xiang Yu was left on the bank speechless in tears.

THE XINAN PONY/SOUTH-WEST PONY
Xinan means south-west in Chinese and includes the provinces of Sichuan, Yunnan, Guangxi and Guizhou, which are where the Xinan pony is mainly distributed. These animals are also sometimes called Guoxia ponies, which translates as ponies under fruit

trees and is derived from the fact that the pony is so small it can be ridden to hack under fruit trees. Guoxia ponies account for 1 per cent of the local equine population.

Fig. 4.4 Following a blizzard, the Hequ and PLA struggle through snow lying over a metre deep and in temperatures of about -18°C.

South-west China's sultry climate is in fact not suitable for horses and the South-west ponies are not thought to be indigenous to the region, but to have evolved from other breeds.

From as far back as the Han dynasty, Korea exported locally bred ponies to China as tribute ponies, although it is unlikely that the limited number of ponies which arrived this way can

account for the entire national pony population in China. The blood-sweating purebred also arrived as tribute from other nations and came in small numbers. China's emperors were very eager to breed it in their own country, but it nevertheless became diluted with a jumble of local breeds. Something similar happened with the Korean ponies.

Historically, the Qiang ethnic group was an ancient nomadic-based tribe which inhabited the western highlands of Qinghai, Gansu and Ningxia. As early as the Xia dynasty (21st–18th centuries B.C.), the Qiangs began to move eastwards and they continued over the centuries to migrate to China's inland provinces with their horses. They finally settled in south-west China, which is hilly and in areas mountainous, lying at an altitude of between about 800 metres in the east to 2,200 metres in the west. The region enjoys a temperate climate with warm summers, mild winters and ample sunlight. These environmental factors mitigate the disadvantage of humidity to the horses. The tribes lived and assimilated with the Han people, and their horses adapted to their new environment and eventually evolved into today's Xinan ponies.

The ponies have relatively large heads, long and thick bristly tails and manes, robust muscle tendons and hard hooves. The hooves are in fact so hard that shoeing them is a problem and they often go unshod. Their narrow chests and sickle-shaped hind legs facilitate their ability to eat and move on steep gradients.

Recognized for their small size, which results in quick and easy heat dissipation during the summer, and above all a low centre of gravity and a higher level of stability, the ponies are especially adapted for travelling in mountainous areas. They can trek along narrow paths and ridges on the very edge of a hanging precipice, and trot or gallop up and downhill. On rough and dangerous terrain, they are agile but highly cautious and sure-footed. The pack ponies can trudge for between 30 and 40

Fig. 4.5 The Hequ has a close resemblance to Emperor Qin Shi Huang's terracotta horses.

Fig. 4.6 Draught ponies carry double wicker panniers across rapids in the Qin Mountains, Shaanxi.

kilometres a day with a load of 100 kg. This unique ability, unusual in a "normal" horse, makes them a very important mode of transportation in rugged areas (Fig. 4.6).

In the vast territory of south-west China, the ponies have become differentiated into various ecological types and varieties. These include the Bose, the Guizhou, the Jianchang, the Lichuan, the Ningqiang and the Yunnan. They are all true ponies.

THE PRZEWALSKI'S HORSE

The Przewalski's horse is the only known surviving wild equine native to the arid desert prairie regions in Central Asia, specifically Outer Mongolia and China.

The critically endangered subspecies of wild horse is named after the Russian geographer and explorer Nikolai Przhevalsky who officially found the herd in 1876 in the Takhiin Shar

Nuruu Mountain area of the Dzungarian Basin in Xinjiang. Many of these horses were captured around 1900 by Carl Hagenbeck, a wild animal merchant, and sold to zoos in Europe. The world population of today's Asiatic "wild" horses is all descended from these "captive" horses. The equid is also called the Mongolian wild horse, the Asian wild horse, the Dzungarian horse and the Takhi (its Mongolian name). Unlike other "wild" horses such as the American Mustang or the Australian Brumby, which are essentially feral horses which have escaped or been turned loose from domesticated stock, the Przewalski's horse has never been domesticated and remains a truly wild animal whether it is on a steppe or in a zoo.

In the Lascaux cave of south-west France, Paleolithic cave paintings abound, dating back to 17,000 B.C. They host nearly 2,000 figures, which can be classified into three main categories:

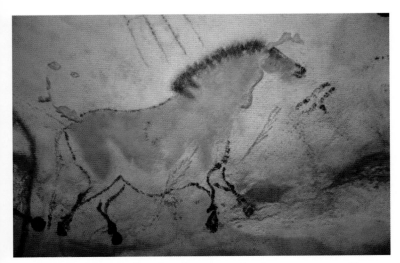

Fig. 4.7 Painting of a dun horse at the Lascaux cave, France. The primitive representation bears a close resemblance to the Przewalski's horse in terms of the classic dun coat and white belly, testifying to the fact that the breed was plentiful in Europe in the prehistoric era.

animals, humans and abstract signs. Of 900 animal images, equines outnumber the rest; they account for 364 depictions. The primitive horse (Fig. 4.7) representations bear a close resemblance to the Przewalski's horse in terms of the classic dun coat and white belly, testifying to the fact that the breed was plentiful in Europe in the prehistoric era.

However, significant changes in ecosystems at the end of the last ice age contributed to their habitat destruction. In the first half of the 20th century, increased hunting and poaching, and competition for grazing with domestic livestock further added to the hazards, sending the species to the brink of extinction.

The Przewalski's horse stands about 13 hands (132 cm) high and, like the Mongolian horse, is still considered a horse not a pony.

As a truly wild horse, the muscular animal bears a number of characteristic primitive markings: dun coat, bold dorsal strip running down the spine from ears to tail, up to 10 black stripes on the back of the lower legs and tail guard hair in coat colour for about 30 cm. The dun coat has a reddish tinge occasionally and finishes with a pale underbelly and muzzle. The animal's large head is topped by a dark and erect mane, with no forelock. The Przewalski's horse sheds its hair all at once, while its domestic relative sheds in a more diffuse and irregular fashion.

Looking into its genetic make-up, the Przewalski's horse possesses 66 chromosomes, two more than its domestic relative. Strangely, a hybrid of horse and donkey, i.e. mule or hinny, is usually sterile, especially the male offspring. It is claimed that the inherited 63 chromosomes of the interbred equid cannot divide into pairs in order to create successful embryos, rendering most mules and hinnies infertile. However, a hybrid of a Przewalski's horse and domestic horse can still breed and produce fertile offspring despite having 65 chromosomes, which is also

Fig. 4.8 A one-month-old Przewalski's foal and its mare in the Beijing Daxing Elk Park. More than 500 Przewalski's horses have been born in China in the last 30 or so years.

Fig. 4.9 The Kalamaili Mountains encompass a vast expanse of territory, which ranges from burning desert to snowy mountains, with scarce forage and water sources available for the Przewalski's horses and other native ungulates that nevertheless find the unique natural landscape to their liking.

an odd number. According to Dr. Ann Chandley, the world-renowned cytogeneticist who has studied hybrids, at the primary spermatocyte stage, horse–donkey hybrids show a "block" which is caused by incompatibility of synaptal pairing between paternal and maternal chromosomes, resulting in a complete arrest of spermatogenesis.

Although the Przewalski's horse hybrid is fertile and looks exactly like the wild parent, it could prove detrimental to the valuable Przewalski's species as it would dilute the precious few Przewalski's genes that are left.

CONSERVATION OF THE PRZEWALSKI'S HORSE

In 1977, the Foundation for the Preservation and Protection of the Przewalski's Horse was established in Rotterdam, the Netherlands. Since 1992, more than 300 wild horses have been reintroduced to their native habitat in Outer Mongolia at the Khustain Nuruu National Park, Takhin Tal Nature Reserve and Khomiin Tal. A reintroduction programme has also taken place in each of Kazakhstan and Ukraine.

In China, the rare species was last seen in the wild in 1966. With the assistance of the International Rare Animal Protection Association, the Przewalski's Horse Reintroduction Programme of China was initiated in 1985 when 11 wild horses from East Germany and England were reintroduced. They stayed first in the Wulumuqi Zoo and then in the following year moved to the Xinjiang Wild Horse Reproduction and Research Centre, which was established to provide conservation facilities for the wild horses. In 1988, the first filly was born and subsequently, 13 more horses were introduced to the centre from overseas. In 2001, the first herd of horses was released to the 1.7-million-hectare Kalamaili Ungulate Nature Reserve, which is also a refuge to a variety of other symbolic species, including the Asiatic wild ass (Khulan), Goitered gazelle, Argali sheep and wolves. Under a dedicated breeding programme, in excess of 500 more foals have arrived (Fig. 4.8), about 60 of which were born in the wild. The International Union for Conservation of Nature and Natural Resources has reclassified the horses from "extinct in the wild" to "endangered".

In early 2020, the population of the truly wild equid was around 2,000, and of these over 600 were in China, including 240 animals roving

and grazing in the Kalamaili Ungulate Nature Reserve. In neighbouring Gansu, 107 animals have been resettled in the Dunhuang Xihu National Nature Reserve, an area which represents another cradle where ancestors of these wild horses once roamed free.

Since 2006, through cooperation with the Cologne Zoo along with the assistance of the Smithsonian National Zoo, horses ready to be released have been collared with GPS receivers, VHF radio beacons and satellite transmitters so that the centre and ecologists can track their movement and develop, implement and monitor conservation strategies.

The reintroduction programme is however not without challenges.

The species lacks genetic diversity. All the Przewalski's horses alive today are offspring of the 14 wild horses captured by Carl Hagenbeck in the 19th century, which means close relatives were forced to breed. Inbreeding results in the Przewalski's horse having lower levels of genetic variation than other equids, leading to low survivorship, poor sperm quality and greater vulnerability to disease. Worse, under long-term confinement, the horse exhibits degeneration of wild instincts in such areas as the teeth and running speed, affecting its ability to acclimatize to the rigours of the Dzungarian Basin's climate, according to Cao Jie, director of the Xinjiang Wild Horse Reproduction and Research Centre.

Habitat degradation is also a concern. The Przewalski's horse had disappeared in the wild for over three decades. Even though the horses have returned to their ancestral geographical range, the ecological environment now is totally different from what it once was. The Kalamaili Mountains encompass a vast expanse of territory ranging from burning desert to snowy mountains, with scarce forage and water sources available for a group of endemic ungulates that nevertheless finds the unique natural landscape to its liking (Fig. 4.9). The winters are so harsh and frigid that a snow layer can develop up to a metre thick, making the search for food more difficult. In addition, global warming has caused the climate to become more extreme in recent years. Released horses are often found dead after frosts or blizzards, despite the provision of relief food and water by the conservation centre. A highway runs through the middle of the reserve, which affects the ecology and environment adversely, and what is even more frustrating is that the horses are also accidentally killed by trucks.

Another inhabitant of the area is the nomadic Turkic Kazak. These people rely on the Kalamaili Mountains for the survival of their herds and their own livelihood. Peaceful co-existence of the nomadic pastoral society and the area's endemic wildlife is difficult. The nomads' 200,000-plus livestock add further pressure on the grassy reserve as they mow across it twice a year. The nomads' domestic horses will probably mate with the Przewalski's horses, whose precious genes will be diluted as a result. Many Kazaks, nevertheless, support the conservation programme, in particular the elders who missed the Przewalski's horses which they knew once roamed in the area. Unlike the Outer Mongolian government, which bought up the relatively small numbers of the nomads' livestock and put them in a reserve, employing the nomads to patrol it, the Chinese government cannot make the same arrangements in the Kalamaili Mountains for the Kazaks as there are far too many of them and their herds. Professor Hu Defu of Beijing Forestry University explains that if the nomads' traditional way of living is to change, they first need to be given an alternative means of existence. Otherwise, it will be the nomads rather than the Przewalski's horses which face a survival crisis.

The road ahead for the Przewalski's horse will be long and daunting, and will require substantial and continual effort.

Fig. 4.10 (Opposite) The Erguna Wetland is the habitat of the Sanhe. The soil of this endless prairie wetland is known to contain plenty of lime, which encourages the growth of good quality forage grass suitable for grazing.

Fig. 4.11 (Overleaf) Sanhe mares and their foals in the Erguna Wetland.

Fig. 4.12 A herd of Sanhe horses crosses a river in the Erguna Wetland, Hulunbuir, Inner Mongolia.

DEVELOPED BREEDS

Most developed breed horses evolved during the early 1900s when China and Russia both underwent political upheavals. As the two countries are geographically close, it was inevitable that when the Bolshevik Revolution broke out in 1917, tens of thousands of Russians would flee into China. Among the exiles were Russian aristocrats and White Russian army officers who came with their horses, and these in time crossed with the native Chinese breeds, cultivating horses like the Sanhe and Yili.

THE SANHE HORSE

Sanhe means three rivers in Chinese. The Root, Drbout and HaBuer Rivers of Erguna in the eastern part of Inner Mongolia drain the area's grassy basin and this has become the Sanhe's habitat (Fig. 4.10). The three rivers form part of the Erguna River (or Argun in Russian), which itself forms part of the border between China and Russia. The region spreads further to Outer Mongolia's San Beise (nowadays Choibalsan) from where the best Sanhe horses were raised according to Henry (Harry) Morriss, son of a Mohawk chief and owner of the *North China Daily News* in the late 19th century.

As the homeland of Genghis Khan (1162–1227), the basin is noted for its continental climate, magnificent steppes and lush wetlands (Fig. 4.11). Although the soils are less suited to arable farming and forest, Henry Morriss discovered that the soil contained a substantial quantity of lime, which can produce high quality forage grass especially suitable for grazing animals (Fig. 4.12).

This region was literally the cradle where horses were domesticated millennia ago. Temperatures in Erguna can plunge as low as -30 to -40°C in winter. Its rigorous conditions have enabled the Sanhe horse to evolve with pronounced greater health, soundness, robustness and stamina compared with horses in Europe where the living environment is more favourable. They can also recognize and avoid poisonous vegetation. Kept in the open, the Sanhe horse adapts to the freezing weather by reserving its body fat during the short summer and by growing a thicker skin and a denser fur coat in the winter.

The pedigree of the Sanhe horse has a long history. Going back to the Liao empire around the year 1000, the region was notable for its quality horses, which were in part made up of tribute horses from nomadic tribes. During the Qing dynasty, some 800 years later, this area saw the creation of the Suolun breed, which was developed to provide cavalry horses for use against invaders.

After the Bolshevik Revolution in 1917 and the ensuing Civil War, thousands of Russians made their way to China together with horses like the Buryat (Zabaikal), Orlov and Bityug (Biçuk in Turkish), which are of a bigger physique and were used to improve the native breed.

The Buryat is a versatile little horse, used for transport, riding and agricultural purposes. The Orlov Trotter is the oldest and most famous breed in Russia. It has a swift and balanced trot, and a reputation for its outstanding speed and stamina. The Bityug horse is raised in the Ukraine region and was developed by crossing the Orlov, Danish and Dutch breeds. It has a handsome appearance, and a good trotting speed and strength. The three breeds helped to foster the Sanhe as a mid-distance horse and sprinter compared with the native Mongolian breed, which is an endurance horse.

From early 1932 through 1945, the Japanese occupied the region and formed the puppet kingdom of Manchukuo. They established a stud farm, bringing Anglo-Arabs, Arabs, Thoroughbreds, American Trotters and Chitran horses to intercross with native varieties. In 1955, after an assessment of the developed breeds in the area, the Ministry of Agriculture of the PRC established two stud farms to develop a new breed and the Sanhe breed finally evolved. It is said that the horse can trace its pedigree to a maximum of 14 breeds.

The horse has a chestnut, bay or black coat, stands between 13.3 and 14.2 hands (135 and 144 cm) in height and weighs between 330 and 380 kg. It is muscular, with sloping shoulders, strong joints and hard hooves. The Sanhe horse possesses a long chest cage and a flat back. As a saddle horse, it can cover a kilometre in 70 seconds; as a draught horse, it can pull a cart with a load of 500 kg.

ANTARCTIC EXPEDITION

The docile Sanhe horse is primarily used in racing, saddle and harness work. In 1907, they were even recruited to explore the South Pole (Fig. 4.13).

Sir Ernest Shackleton took 10 of these small horses on his first expedition to Antarctica. He praised them, saying: "…compared with the dog, the pony is a far more efficient animal, one pony doing the work of at least ten dogs on the food allowance for ten dogs, and travelling a further distance in a day. […] It was trying work for the ponies, but they all did splendidly in their own particular way."

Fig. 4.13 In 1907, 10 Sanhe horses were recruited by Sir Ernest Shackleton to explore the South Pole. Two did not survive the ocean voyage, four died from eating volcanic sand on arrival and the remaining four – Socks, Grisi, Chinaman and Quan – all died in the expedition southward to the Pole.

Shackleton followed the British Army's advice to feed the horses with a meat-based supplement which was called "Maujee ration" and consisted of dried beef, carrots, milk, currants and sugar. Horses are herbivores, but it was found that they preferred Maujee ration to the traditional fodder given on the polar journey.

Sadly, two did not survive the outgoing ocean voyage, four died from eating volcanic sand for its salt content upon arrival and the remaining four – Socks, Grisi, Chinaman and Quan – all died in the expedition southward to the Pole. The horses, nevertheless, still made history. Socks was the last surviving horse. Shackleton heard it mourning all night over its deceased friends. It finally fell into a bottomless crevasse just hours before it was due to be shot to feed the crew.

Although Shackleton and his toughest horses both fell short of the goal to cross Antarctica, Socks was the equine which has come closest to the South Pole in history.

RACING CHAMPIONSHIPS

The Sanhe horse is the only home-grown racing horse in China. The breed was very active across China and had raced in Hong Kong since 1856. Sanhe horses new to the tracks were called griffins, a term which was used to denote newcomers to India, or novices. The crossbred was developed in the early 20th century and was faster than ever. Liberty Bay and Silkylight were among the greatest Sanhe in Hong Kong's racing history. Both were crossbreds. Liberty Bay won the Hong Kong Champions three years in a row, always ridden by Leo Frost. Six lengths were the usual margins of its victories. It held a track record of 26 wins and had never known defeat until the appearance of Silkylight.

Silkylight was of Shanghai origins. The black horse was injured and found by Eric Moller when bombs dropped by the Chinese air force failed to hit Japanese cruisers but killed and wounded thousands of civilians in Shanghai in August 1937. Although it did not impress when it debuted in Hong Kong, it went on to win the Maiden Stakes and then the Derby where the winning ticket yielded an enormous HK$141,380 purse (US$18,126). Exciting enough, if it further won the Hong Kong Champions, it would sweep the triple crown. On February 24th 1938, while Liberty Bay was still the hot favourite with its invincible records, to everyone's surprise, Silkylight beat the unbeatable and won the Hong Kong Champions. The race was so sensational that everyone went on to remember it in the history of Hong Kong racing. Lambert Dunbar, owner of Liberty Bay, was so upset that he never allowed the horse to race again. In November 1938, Silkylight won the Hong Kong St Leger in a record 3 minutes 29 and three-fifths seconds, the best time by a Sanhe in China.

Although horse racing completely disappeared in China when the Communist Party came to power in 1949, racing continues in Hong Kong, but since 1971 and 1972, only Thoroughbreds have been allowed to race. The Sanhe had passed its prime.

Horse farming is not profitable in modern China. The breed has been on the verge of extinction and many of these horses have been sold to Japan as food. In early 2000, about 3,000 animals remained, which is one-tenth of the population 10 years earlier. In recent years, a Sanhe conservation campaign has been launched by creating a more economically viable

Fig. 4.14 The Yili is thought to be a cross between the Nisaean from Fergana and the Mongolian-type horse. Emperor Han Wudi first gave the breed the name heavenly horse but later renamed it western extremity horse, using the name heavenly horse for his favourite Nisaean horses instead.

model of the breeding business in order to maintain its survival. This includes bundling breeding with racing, riding school, equestrian museum and tourist attractions. It is expected to see the breed thriving in the area around the Erguna River again in the foreseeable future.

THE YILI HORSE

The Yili horse (Fig. 4.14) originated in the Yili–Kazakh Autonomous District of Xinjiang. In ancient China, Yili was known as Wusun and was the home of the prized Wusun horse. The Wusun is thought to be a cross between the famed blood-sweating purebred from Fergana and the Mongolian-type horse. It was first obtained by China in 115 B.C. during the reign of Emperor Han Wudi (reigned 140–87 B.C.) who had a princess married off into Wusun to cement a bond between the two royal houses. Han Wudi first gave the Wusun horse the name heavenly horse but later renamed the breed western extremity horse, using the name heavenly horse for his favourite Fergana horses instead.

The Yili horse is raised in the counties of Zhaosu, Tekesi, Xinyuan, Nileke and Gongliu, which feature a temperate, continental and arid climate. The area is characterized by snow-capped peaks, charming highland lakes, lush meadows and dense forests, with steppes and valleys scattered between them. Rich natural and mineral resources abound (Fig. 4.15).

Fig. 4.15 A herd of Yili horses runs in Zhaosu pasture.

In the early 1900s, the Yili breed evolved from the cross-breeding of the Russian Don, Orlov Trotter and Anglo-Don as the paternal lines, with the native Wusun horse as the maternal line.

Considered the ideal horse by millions of Cossacks, the Russian Don is known for its endurance and stamina. The Orlov Trotter is one of Russia's most highly respected breeds, famed for its tremendous trotting power and stamina. The Anglo-Don is a breed obtained by crossing the Thoroughbred and Russian Don. Most Yili horses have at least 50 per cent Russian Don blood in them.

The Yili is a light riding and driving breed. It has an average height of 14.1 to 14.2 hands (143 to 144 cm) and weighs 400 to 450 kg. The horse's compact and harmonious conformation comes with a light head, big and bright eyes, long neck and tough limbs. With the nape of its neck held high

Fig. 4.16 Khanbegler, champion of the 2011 Akhal-Teke beauty pageant in Turkmenistan. The country has declared the pedigree a national emblem.

and a body coat of bay, chestnut or black, it looks mighty and imposing, but it can also appear delicate and graceful. Gentle, agile and good at hopping over obstacles, the Yili treads safely on mountainous terrains as well as plains. It can carry an 80-kg load to finish 126 kilometres in just over seven hours, proving its unequalled stamina and hardiness.

The Yili represents 60 per cent of the sport horse population in China, the breed possessing pronounced sporting ability, especially in endurance. Not only does it boast a refined appearance, but many local riders claim the Yili as their favourite mount. In December 1983, two Yili horses were chosen by the Chinese government as diplomatic gifts to the late King Hassan II of Morocco, who later bought some more of the breed from China.

EXOTIC BREEDS

The Blood-sweating Purebred
The name of this horse arose from the legend that the animal produced sweat the colour of blood during exertion.

It is probable that the blood-sweating purebreds were the early ancestors of the modern Akhal-Teke breed (Fig. 4.16). The horse was first officially mentioned in Chinese history in 126 B.C. when Zhang Qian reported to Emperor Han Wudi that in Fergana (modern day Uzbekistan) he had encountered amazingly fine horses that sweated blood.

Earlier in history, during the fourth century B.C., when Alexander the Great had marched to Margiana in southern Turkmenistan, he met fierce resistance from nomadic Sacae and Massagetae tribes. Descriptions of their horses were not very dissimilar to those reported by Zhang Qian.

Emperor Wudi was absorbed with the idea of obtaining the blood-sweating purebred because of China's chronic need for high quality horses and his belief that the horses might be the legendary heavenly horses which could carry him to the home of the immortals.

In order to secure the horses, Wudi launched several military expeditions against Fergana, decimating the tens-of-thousands-strong Han army and in the process expending an enormous amount of resources.

Modern authorities attribute the blood-coloured sweat to a parasite that is picked up in the rivers of the Gorgan and Fergana region. At a certain stage in its lifecycle, the parasite breaks through the host's skin causing bleeding. This bleeding also occurs in other animals such as donkeys and cattle, but not human beings.

Some had considered that the colour was an optical illusion on the reddish-coated horses. Equine sweat has such a high concentration of protein that it will foam when agitated by working muscles. During intense exercise, the horse will lather up in areas where sweat glands are prolific: the neck, shoulders and between the pelvic limbs, and the reddishness of the purebred was thus accentuated, creating an illusion of blood sweating.

BLOOD-SWEATING PUREBREDS IN LEGEND
Throughout the imperial period, China secured a continuing stock of blood-sweating horses through purchase, tribute and plunder both for the emperors' pleasure and in particular for the improvement of local breeds. However, despite the intention, the stock was not widely bred and it did not improve the overall quality of native Chinese animals. This failure can be attributed to China's inability to develop a consistent internal breeding programme and that the stock of blood-sweating horses was still far outnumbered by native breeds.

However, the horses enjoyed unrivalled care at the royal mews. Some became favourites or

Fig. 4.17 The stallion Liman exhibits a glowing coat.

chief mounts of emperors and nobles; some became motifs of great literature and artwork; while others were used in exotic recreational and cultural activities at imperial courts. Their fame is a glorious relic of Chinese history and their tales remain alive and honoured.

Red Hare was a legendary steed in the epic Chinese novel *Romance of the Three Kingdoms*. The

horse was described as being red all over like glowing charcoal – without a single hair of another colour. As for the name "Hare", according to the writings by Bole, the famous tamer and judge of

horses' physiognomy from the Spring and Autumn period, the hare-like head is a fine horse feature. Red Hare was said to be able to cover a thousand *li*, which is roughly 416 kilometres in modern day measurement, in a day – although we know that in reality this is impossible. However, from the horse's unparalleled endurance and appearance, not least its glowing coat (Fig. 4.17), which embodied the blood-sweating purebred's distinct metallic sheen (Fig. 4.18), it is likely that Red Hare was first and foremost one of the superlative "heavenly horses" from Fergana. Throughout the novel, Red Hare was repeatedly used as a treasured gift by the several aggressive political figures with ambitions of being emperor. In their efforts to win favour, Red Hare was basically used as a bribe and during its lifetime, it therefore served several famous owners. These ranged from Dong Zhuo to Lu Bu to Cao Cao to Guan Yu and finally to Ma Zhong. It is said that

the horse refused to submit to this last named and starved to death following the execution of Guan Yu, the Saint of War known to every Chinese household. Red Hare together with Green-Dragon Sabre have become such iconic symbols of Guan Yu that they are now firmly immortalized in the minds of readers.

THE NISAEAN HORSE

The blood-sweating purebreds were thought actually to be Nisaean horses (Fig. 4.19) and originated from the Nisaea plain where the Persian royal stud farms were established. The Nisaea plain is believed to be located in the oasis of Akhal in modern day Turkmenistan.

These horses fed on Median grass, which as previously described in chapter one is a clover that grew in the plain and was a highly nutritious legume fodder with a protein level twice that of most grass hays. The ready availability of this food plant together with selective breeding enabled Nisaean horses to evolve to become the fine animals they were reputed to be.

With a withers height of 150 cm, which was relatively tall for pre-19th century equines, Nisaean horses had attained a robust skeleton and musculature. They were famed for their superior size, magnificence, fine proportions and stamina, though not for their speed.

The Chinese were not alone in favouring these horses; the Greeks (mainly the Spartans), the Scythians, the Assyrians and the Romans all coveted Nisaean horses. They imported, captured, stole, or demanded them as a tribute from the Persian Empire. According to the Roman poet Grattius Faliscus, every aristocrat fancied a Nisaean horse for warfare and hunting. Despite this, Chinese Emperor Shizu Fu Jian (338–385) of the Former Qin once declined this highly sought-after gift and returned the tribute horses including the Nisaean from Fergana in order to keep to his anti-extravagance campaign.

Fig. 4.18 The Akhal-Teke has a deep iridescent, metallic sheen.

Fig. 4.19 A Sassanian rock relief illustrates Ardashir I (left) receiving the ring of power from Ahura Mazda at Naqsh-e Rostam. The emperor and the Zoroastrian god each mounts a Nisaean horse.

In 580 B.C., Cyrus the Great ascended the throne of Persia. The father of the "Iranian nation" made the Nisaean the imperial horse and dedicated it to the Persian deities Ahura Mazda, the supreme god, and Mithra, the god of light, and sacrificed white Nisaeans to Mithra on his December 25th birthday and on New Year's Day. On one occasion, he was so distraught that he drained a river in which his beloved Nisaean stallion had drowned.

Nisaean horse breeding was highly selective. Roman author and naturalist Pliny the Elder (c. 23–79) reported that only those stallions which had won races were chosen for mating.

Nisaean horses were not commonly found in the Roman Empire until the Byzantine era when the horses were bestowed on various Roman emperors.

In 330, Constantine the Great moved the capital of the Roman Empire from Rome to Constantinople. He became fascinated with the Persian cavalry and, based on their military practice,

divided his own Roman cavalry into two different types: the cataphract and the *clibanarii*, though it is now unclear which wore the heavier armour.

Emperor Justinian (483–567) is considered one of the greatest Byzantine emperors. Ambitious to reunite the Roman Empire and reclaim the loss of the historical territory, he set up an imperial stud in Bythnia for Nisaean horses to strengthen his military power. He also began a policy of rewarding Arabs with Nisaean horses for killing Persians while Persians paid back in kind.

In 1204, crusaders invaded Constantinople rather than Egypt, which was their original target in their attempt to capture Muslim-controlled Jerusalem. The Fourth Crusade resulted in the sack of Constantinople by the crusaders and Venetian forces. In the process, the crusaders, in particular the Normans, caused serious destruction to the imperial Byzantine stud. They sought to refine their own horses by crossing them with the Nisaeans:

the French Boulonnaises and Percherons were thus improved. Nisaeans were subsequently dispersed throughout Europe, leading to the creation of the Barb and Andalusian. After this scattering of the stock, Byzantine-bred Nisaean horses were believed to have become extinct.

However, when the Turks conquered Constantinople in 1453, Nisaeans still appeared in Turkish art, which depicted Turkish heroes riding leopard Appaloosas. These were dark-spotted Nisaeans claimed by the Persians to have originated from their legendary war horse Rakush whose unusual, patterned burnt-orange coat was held sacred and likened to "rose leaves scattered on a saffron ground".

Back in Persia, although the Medes still kept great herds of Nisaean horses, the steppe was also inhabited by many other breeds which would mate with the Nisaean. On top of this, the region saw incessant warfare for centuries, making it even more difficult to maintain the horse's genetic purity.

The Persians were obsessive about their sacred horses. If the horses were smuggled out of Persia, Persian vets were forbidden to treat them. Since the horses were raised and maintained with special care, which was not widely understood, the horses' life expectancy was impacted if proper husbandry was not provided for them.

Another Persian tradition required Persian cavalry units to kill their horses if they were defeated so that the horses would not be confiscated by their conquerors. The custom increased the rate of the horse's demise.

Nisaeans are hardly traceable nowadays. However, some Iranian tribes claim that the purebreds still exist, especially around Turkmen Sahra but in very small numbers. The present-day Turanian and Akhal-Teke breeds are considered their direct descendants.

THE AKHAL-TEKE

The Akhal-Teke is an ancient desert breed from Turkmenistan, where they are a national emblem and are treated as family members by their owners. They are noted for their incredible stamina and fortitude, characteristics which are attributed to their small-portioned but high-protein diet that often includes butter and eggs mixed with barley. With superb natural gaits, these horses are especially outstanding in dressage and endurance.

The Akhal-Teke is commonly dun in colour, although it can also be black, bay and grey, with a metallic glow to the coat. The horses stand 15 to 16 hands (152 to 163 cm) high and weigh between 400 and 500 kg.

These "golden-horses" are native to harsh terrains and live under extreme climatic conditions. They possess a quiet temperament, but are easily aroused. They are courageous, alert and intelligent animals.

The global population of Akhal-Tekes is about 3,500 and is found mainly in Turkmenistan and Russia. The Akhal-Teke is among the most elegant of the world's horses and has a distinctive conformation which is sometimes favourably compared to the Persian Arab. It has a fine head and large expressive eyes. The long back is lightly muscled, and is coupled to a long, straight and often thin neck, which is set high onto excellent sloping shoulders and is topped with a short silky mane. The limbs are long and slender with clearly revealed tendons.

In 2002, the late Turkmenistan President Saparmurat Atayevich Niyazov gave an Akhal-Teke horse to China as a gift of diplomacy after President Jiang Zemin visited his country. In 2006, Niyazov brought another Akhal-Teke with him when he visited China. Unlike their ancestors which trekked the length of the Silk Road from Turkmenistan, these horses flew all the way.

In recent years, the Akhal-Teke has been introduced into China by businessmen both for the purpose of breeding and for sales to the nouveau riche, who see the animal with its desert origins and strong imperial background as an exotic pet. An Akhal-Teke imported from Turkmenistan costs at least US$300,000 and is still in high demand.

RECORDS SET BY THE AKHAL-TEKE
In 1935, when Turkmenistan was one of the constituent republics of the Soviet Union, 15

Fig. 4.20 Silent Witness was one of the most celebrated and accomplished Thoroughbred race horses in Hong Kong with 17 consecutive wins.

Akhal-Tekes travelled 3,000 kilometres from Ashgabat to Moscow across the Karakum Desert on a forced march in 84 days, which included going three days without water. This march was meant to demonstrate to Joseph Stalin the breed's formidable strength in the hope that he would give up their annihilation and permit their continued breeding. The campaign was a success.

However, Russians argue that Western media do not understand the political reality in the USSR at that time. The march was just a carefully planned and systematically executed propaganda activity, with the aim of exhibiting the achievements of the Turkmen Soviet Socialist Republic. When the riders arrived in Moscow, a reception was held at the Kremlin in their honour and they were given state prizes. Without the prior endorsement of the authorities, this could not have taken place.

The event nonetheless drew the government's attention to the magnificent native assets of the Turkmen Soviet Socialist Republic, leading to the establishment of new breeding farms and provision of research funds especially for the Akhal-Teke and Yomud.

After World War II, there was a renaissance of the Akhal-Teke breed; the Ashgabat and Djambul studs were set up at that time and produced premium horses.

However, with military modernization, cavalry mounts were gradually phased out and from 1953, many stud farms were forced to close. The breeding stock at Ashgabat was slashed while all of the horses at Djambul were on the point of being slaughtered. Luckily, with the help of Veniamin Veniaminovich Ivanov, the director of the Lugovskoy stud, the horses were rescued and transferred to the modern day Kazakh Republic where Absent was raised and trained to become a world champion.

Absent was a raven Akhal-Teke with four white socks and a star on its forehead. It won the Prix de Dressage at the Rome Olympics in 1960, the bronze individual medal in the Tokyo Olympics in 1964 and the Soviet team silver medal in the Mexico City Olympics in 1968.

Absent gained an aggregate score of 1070 points from three judges in the ride-off test in the 1960 Rome Olympics. The score was among the highest in Olympic dressage history, but actually is not comparable with those of other years as score requirements vary by year, in particular those in 1960 were less demanding than nowadays. In addition, a percentage point system is employed at present; it is impossible to trace back the full marks of the test in 1960 and a percentage point for its score thus cannot be derived.

No matter what, Absent's performance was and always will be seen as terrific.

THE THOROUGHBRED

Byerly Turk was one of the three horses which created the modern Thoroughbred horse-racing bloodstock (Fig. 4.20). While the animal is commonly believed to be an Arabian horse, there is considerable speculation that it was in fact a Turkoman horse or an Akhal-Teke.

Nowadays, only a few modern Thoroughbreds can trace their sire line back to Byerly Turk and one is the Hong Kong-trained gelding Cape of Good Hope, which won the 2005 Golden Jubilee Stakes at Royal Ascot.

Another Thoroughbred foundation sire Darley Arabian, according to Russian hippologist Professor V.O. Vitt, was also thought to be a Turkoman horse or possibly a Turkoman-Arabian cross. In 1704, Darley Arabian was shipped to England from Aleppo in Syria, which was part of the Ottoman Empire from 1516 to the end of World War I. Carsten Niebuhr, a German mathematician, cartographer and explorer who participated in the 1761 Danish Arabia Expedition, indicated that Ottoman Turks had a penchant for Turkoman horses over Arabian horses to which they had

easy access. They liked to decorate the horses with ornate and embellished jewellery.

The Turkoman horses were bred across the Near East, and evidence shows that the breeding continued in Syria until the 19th century. In the 1890s, Dr. O.A. Balakshin, a renowned Russian expert in Arab horse breeding, recognized that the Syrian Arab and the Akhal-Teke share certain similarities in their conformation. Meanwhile, a good many Oriental horses which contributed to the creation of the English Thoroughbred bore a strong resemblance to the Turkoman horse.

Last but not least, the traditional Turkoman practices such as exercising horses in blankets and early breaking and training young stock are also found in English Thoroughbred training. It is quite likely that the practices were brought to England by the grooms and handlers of the horses.

THE FUTURE FOR CHINA'S HORSES

Throughout the imperial history of China, horses were mainly bred for military purposes. After World War II, economic development was of first priority and the horses were reassigned to put their strength behind agriculture and transportation, helping to build up China's economy and to support its people's livelihoods. In recent decades, with the country's emphasis on modernization and urbanization, the value of the horse has plunged.

The net result of this is that the current status of the Chinese breeds described above is not dissimilar to that of their Western counterparts. Most nowadays in the West are seen only in the sports arena where modern equestrian pursuits include horse racing, show jumping, dressage and cross-country.

Horse racing is dominated by the Thoroughbred, which is a hot-blooded horse, known for its agility, speed, boldness and spirit. An ideal Thoroughbred should have a fine chiselled head sitting on a long and light neck, a straight profile, high withers, a deep chest, a short back with good depth in the hindquarters, a sleek body and long legs.

Most breeds of horse have some degree of jumping ability, but some just do not enjoy the

experience. An ideal show-jumper has long lines, a long sloping shoulder, pronounced withers, a well-muscled croup obscuring the bones, clearly visible tendons, correctly positioned legs as well as strong-angled joints. Height does not really matter as there is no direct relationship between size and jumping ability. Most show jumpers nevertheless tend to be tall, over 160 cm, and are usually warm bloods.

German Holsteins, Hanoverians and Polish Anglo-Arabs all exhibit the above qualities and are renowned for their jumping ability. English Thoroughbreds are also good jumpers but they need riders endowed with a great deal of sensitivity as they have temperaments which are often difficult to handle.

Like show jumpers, most dressage champions are warm bloods, although any breed can benefit from dressage training. An ideal dressage horse should appear graceful and noble, preserving an impression of sensuous beauty. It has a moderately long and straight head, large and calm eyes, a long neck and sloping shoulders. The back should be long, strong and muscular, while the legs should be clean and properly positioned, with tendons clearly defined. All movements should be full of energy and elegant. The horse should be high-spirited, but at the same time reserved. The English Thoroughbred's temperament makes it less suited to be a dressage horse.

In addition to the show-jumper's qualities, a good cross-country horse must also have a strong personality and tremendous stamina to be able to tackle the rigours of a cross-country course. First class eventers usually have a high proportion of noble blood pulsating in their veins. Some are pure Thoroughbreds and most carry at least 50 per cent Thoroughbred blood.

No matter which disciplines they undertake, competition horses should enjoy the work, the rivalry with their peers and the exciting atmosphere, but still remain sensitive and calm in their basic natures.

Where into this then do the horses of China fit? As we can see, the native and developed breeds which exist in China today are not qualified to compete in any of the above competitive fields at an international level. Even worse, their value in other areas of Chinese life is something people are neither aware of nor need.

In 1999, the Ministry of Agriculture confirmed that the Elunchun horse (Fig. 4.21), the Tieling draught horse and the Jinzhou horse (Fig. 4.22) were at risk, while the Tengchong (Yunnan) pony, the Jinjiang horse, the Shandan horse, the Heihe horse, the Heilongjiang horse, the Guanzhong horse, the Ningqiang pony and the Yiwu horse were endangered.

On August 23rd 2000, the Ministry of Agriculture announced that the Bose pony and the Mongolian horse were two of the 78 key conservation animal breeds then recognized at state level.

An improvement in the situation is nowhere in sight. In recent years, the demand for both native and developed breeds has decreased even more. The Sanhe horse, the Xilinguole horse and the Yili horse have joined the declining group.

Looking to the future, we must ask whether the Chinese breeds can reinvent themselves from war and draught horses to sport horses. Modern equestrian sports and sport horses alike were developed by Western countries over a period of more than a hundred years. China lags very much behind in knowledge, culture and talents – both in terms of horses and humans. It will be a challenge for both the nation and the Chinese breeds to catch up with their Western counterparts.

Today, the Chinese breeds, which have been developed over the centuries to stand rigorous conditions, vast open spaces and meagre feeding, and to undertake military, agricultural and transportation missions, have problems simply surviving the changing economic and social environment. As they are no longer needed as helpers, no practical value is seen in them. Should they then just be left to die out and be lost to time?

As a part of Chinese history, they deserve recognition and every accolade for all they have contributed to the civilization, and they warrant conservation of their unique genetic diversity. Rather than being a challenge to the horses, China's endangered horses' survival relies more on the awareness of the nation. As always, the horses' destiny will be determined only by the action of humans.

CHAPTER FIVE

TRADITIONAL CHINESE EQUINE VETERINARY MEDICINE

INTRODUCTION

Traditional Chinese medicine (TCM) is an ancient art and a natural science. From its origins in China some 3,000 years ago, it has developed over the millennia and is still practised today.

The philosophy of TCM is rooted in the treatise called the *Yi-Jing* (Classic of Change) that includes theories of Yin and Yang, the Five Phases and Taoism (Fig. 5.1).

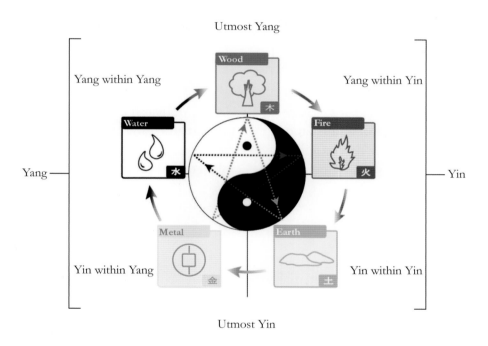

Fig. 5.1 Theories of Five Phases and Yin and Yang.

These theories are based on treating the body as an organic whole, rather than just treating the disease or signs. Traditional Chinese medicine takes into consideration a person's physical, spiritual, social and mental well-being. According to a Chinese saying, "Chinese medicine treats humans, while Western medicine treats diseases."

The same theoretical concepts are applied to animals and have been developed in horses as traditional Chinese veterinary medicine (TCVM).

Legend has it that Ma Sihuang was a renowned equine vet during the Huang Di period (3000–2357 B.C.). His fame spread far and wide, and even to heaven from where an ailing dragon made a special trip to see him. Ma Sihuang performed acupuncture on the dragon's tongue and lower lip and provided it with a serving of herbal tea. The dragon eventually recovered from its ailments, and as appreciation it submitted to Ma who then rode it back to heaven.

Oracle bone scripts are among the earliest Chinese writings and date back to the Shang dynasty (1766–1046 B.C.). The majority refer to pyromantic divinations (prediction by applying fire to inscribed animal bones or shells) of the royal house of the Shang dynasty, but plenty of TCVM practices such as neutering and acupuncture on horses and cattle were also recorded.

Sun Yang (680–610 B.C.) was a legendary horse expert during the Spring and Autumn period. He was so talented at judging and selecting horses that everyone called him Bole, which is the name of a celestial body fabled to be in charge of heavenly steeds. His book, *Physiognomy of the Horse*, is the oldest work about horses and horse husbandry in China. It was lost long ago, but its contents have been so often quoted by other surviving treatises to indirectly prove the book's existence.

Also competent in treating equine diseases and familiar with the equine meridians (see page 131), Bole was credited with being the pioneer of veterinary acupuncture.

Although horses are much like other livestock and were traded in the market, they were considered of higher value than other animals in ancient China. Horses were a major force used in warfare and were held in close association with man on the battlefield. Real horses or replicas were buried with their owners, accompanying them into the afterlife. As horses were considered such important livestock animals, equine veterinary skills became especially sophisticated and a main focus of TCVM. Having said that, the principles and practices of TCVM are applicable to all types of animals and textual references specifically to the horse are not very extensive. But it was for the horse that the practices were first developed and have been passed down.

The *Huang Di Nei Jing* or *Nei Jing* (Yellow Emperor's Classic of Internal Medicine) is known as the earliest extant treatise which systematically describes all the theories of TCM. It is believed that the book was written during the Warring States period (475–221 B.C.). Other important equine reference books include *Simu's Collection of Equine Medicine* edited by Li Shi c. 838 during the Tang dynasty and *Yuan Heng's Collection of Horse Treatments* by veterinarian brothers Yu Benyuan and Yu Benheng in 1608 during the Ming dynasty.

YIN AND YANG

In TCM, health is believed to come from a balance of Yin and Yang (Fig. 5.2). These two opposing yet complementary forces are found in all non-static objects and processes in the universe.

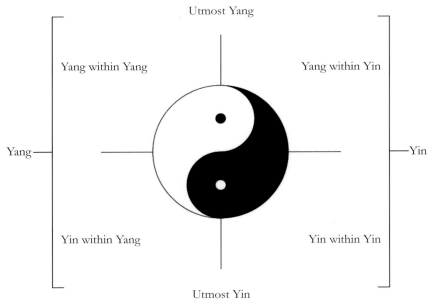

Fig. 5.2 Yin and Yang.

Yang represents energetic qualities like movement, heat, brightness, stimulation and activity. Yin, on the other hand, represents less energetic qualities such as rest, cold, darkness, condensation and inhibition. The balance of Yin and Yang is neither static nor absolute, even in a state of well-being. While galloping, a horse is excited, demonstrating Yang, and yet once it slows to a walk, it becomes relaxed, and Yin dominates. However, when the balance is consistently disturbed and one force regularly dominates the other, then health is harmed and illness can result.

The two forces are in constant motion, maintaining a dynamic and relative balance between each other, and ensuring vitality of the body.

FIVE PHASES

The Five Phases theory of TCVM is known as *Wu Xing* in Chinese. Wu means five and the theory categorizes substances and phenomena of nature into the Five Phases, i.e. Metal, Wood, Water, Fire and Earth. Xing means functions and changes. The Five Phases have their own characteristics, which are inter-related and active, rather than isolated and dormant. They are mutually and constantly generating and overcoming in order to maintain harmony and balance, and activating growth and development of natural phenomena.

In TCVM, the theory is applied to the dynamic equilibrium within the body–mind–spirit of an animal. It explains features and associations of different organs in the animal's body and provides a clinical guide to diagnosis and treatment.

Wood relates to growth, rise and stretch. Fire represents burning, heat and upward movement. Earth symbolizes absorption and transformation. Metal means refinement, reserve and clearance. Water epitomizes downward movement, moisture and chill.

Table 1. The Dynamic Associations of the Five Phases

	The Five Phases Associations				
	Wood	**Fire**	**Earth**	**Metal**	**Water**
Yin Organs (Zang)	Liver	Heart	Spleen	Lungs	Kidneys
Yang Organs (Fu)	Gallbladder	Small intestine	Stomach	Large intestine	Urinary bladder
Sense Organs	Eyes	Tongue	Mouth	Nose	Ears
Seasons	Spring	Summer	Late summer	Autumn	Winter
Climatic Conditions	Wind	Summer heat	Dampness	Dryness	Cold
Tissues	Tendons	Vessels	Muscles	Skin/Hair	Bones
Emotions	Anger	Joy	Pensiveness	Grief	Fear
Colours	Green	Red	Yellow	White	Black
Taste	Sour	Bitter	Sweet	Pungent	Salty
Voice	Shout	Laugh	Sing	Cry	Groan

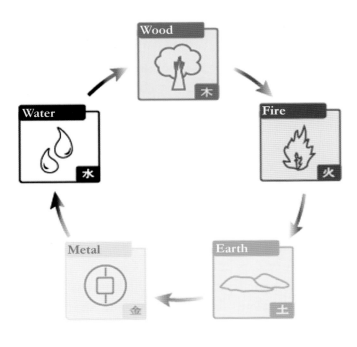

Fig. 5.3 Generating sequence.

THE INTERACTION AND RELATIONSHIP BETWEEN THE FIVE PHASES

The Five Phases all exist in a horse in a balanced manner. If any one of them is missing or insufficient, or alternatively dominates, there will be a negative and profound impact on the horse's vitality. Their properties and inter-relationships with the body can be summarized in Table 1.

The three most common interactions and relationships between the Five Phases are the generating sequence, the overcoming sequence and the balancing sequence. These are explained below.

GENERATING

"Generate" is translated from the Chinese word *sheng*, which actually carries several other meanings in this context, such as create, promote, nurture, support and enhance. Each phase generates the other and the sequence repeats itself in a cyclical manner, giving rise to growth and nourishment (Fig. 5.3).

Wood feeds Fire → Fire creates ash which forms Earth → Earth bears Metal → Rivers and springs rich in minerals are much valued and it is thus induced that Metal promotes Water → Water nourishes Wood.

The relationship can be understood as a child and its mother. For example, Wood is the child of Water and the mother of Fire. The principle can also be applied to the visceral organs which correspond to the Five Phases. For example, the spleen, which is associated with the Earth phase, is the mother of the lungs, which are associated with the Metal phase.

Physiological functions of the lungs are fuelled by food and water essence derived from the spleen. Spleen deficiency will lead to water retention, finally producing phlegm. In a serious situation, phlegm obstruction results, upsetting the disseminating and descending functions of the lungs, thus bringing about a stuffy nose, panting, sneezing and absence of perspiration, among other things.

OVERCOMING

"Overcome" is translated from the Chinese word *ke*, which also means control and eliminate. The sequence can help maintain a self-regulating balance in growth and development of natural phenomena (Fig. 5.4).

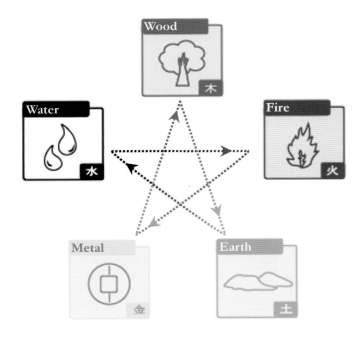

Fig. 5.4 Overcoming sequence.

Wood depletes the nutrients of Earth → Earth absorbs Water → Water extinguishes Fire → Fire melts Metal → Metal chops Wood.

In TCVM, the spleen, which is associated with the Earth phase, is responsible for absorbing and transporting aqueous fluids in order to avoid flooding the kidneys, which are associated with the Water phase, and causing fluid retention. The liver, which is associated with the Wood phase, dredges and drains to ensure the proper flow of qi (see below), thus resolving any congestion in the spleen and stomach, which are associated with the Earth phase.

BALANCING

The balancing sequence is the combination of the generating sequence and overcoming sequence (Figs. 5.5 and 5.6). Promotion and support are provided to the Five Phases to prevent restriction and suppression. Each phase interweaves with the other four. For instance, on the one hand, Fire generates Earth, and Wood generates Fire. On the other hand, Fire overcomes Metal but is

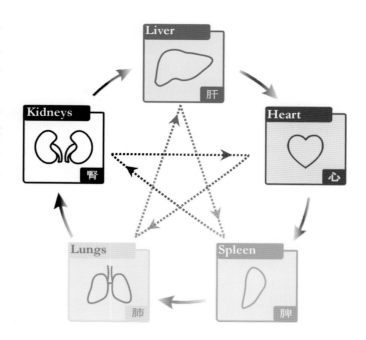

Fig. 5.6 Balancing sequence in zang organs.

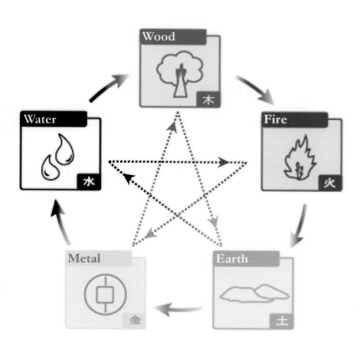

Fig. 5.5 Balancing sequence.

itself overcome by Water. Without the generating sequence, growth and development will stagnate; without the overcoming sequence, balance and coordination of different substances and functions will be impaired. The interaction establishes a dynamic equilibrium in nature.

QI, BLOOD AND FLUIDS

Qi, blood and fluids are the fundamental substances of the horse's body and all its natural physiological processes.

The growth and metabolism of these substances depend on the functions of the *zangfu* organs and meridians, which themselves rely on the nutrients provided by qi, blood and fluids to function properly. In this way, qi, blood and fluids are literally both the products and the energy of the zangfu organs and meridians.

Qi

Qi is loosely translated as "vital energy", "life force", or "breath", but no single English word or phrase suffices to explain it properly. It enables humans and animals to function, grow and reproduce. It also defends the body against illness, warms the body and keeps the organs, tissues and substances in their appropriate places. Qi transforms substances in the body into essence or vital energy, the process of which is imperative for the metabolism of fundamental substances. There are two main sources of qi: one is prenatal and evolves from the biological parents at conception; the other is postnatal and evolves after birth and throughout life from essential substances in nature, such as the air we breathe and the food we eat.

Other than as the fundamental substances of the body, qi itself has multiple meanings. The six evils refer to evil qi. Yang qi pertains to substances or functions with superficial, ascending, exuberant, light and functional attributes. It is the opposite of Yin qi. Healthy qi is understood to mean normal physiological function and disease resistance in the body. It is needed to maintain the well-being of the body, including its adaptability, immunity and recoverability. The physiological functions of visceral organs are also known as qi. There is heart qi, kidney qi, liver qi, lung qi, spleen qi, among others. Whatever qi is attached to, the combination gives qi a different meaning from every other combination.

Blood

The principles of TCVM hold that blood is an essential and nutrient red liquid in the body. Propelled by qi, blood circulates in vessels around the body. It provides nutrients to the internal organs and also to the skin, flesh, bones and muscles, delivers clearing qi to every part of the body and carries turbid qi to be eliminated by excretion, as well as supporting mental activities through nourishing *shen*, which is the root of the spirit.

Body Fluids

Body fluids are the different physiological fluids of the body. They include fluids and secretions from the organs, gastric juice and intestinal juice, nasal discharge and tears.

Body fluids provide warmth and moisture while nourishing different parts of the body. In this way, the balance of Yin and Yang is regulated and metabolic waste and toxins are excreted from the body, leading to a clean and healthy environment for the organs, meridians and tissues.

ZANGFU

The zangfu theory describes the functions and interactions of the 11 organs in the body and proposes that generalized symptoms and signs are the direct expression of the general physiological and pathological state of the internal organs.

Zang refers to the Yin organs, while fu refers to the Yang organs (Fig. 5.7). Each Yin organ is paired with a Yang organ, and each pair corresponds to one of the Five Phases. For example, the lungs and the large intestine are Yin and Yang Metal organs, respectively, and they resonate with each other.

Zang includes the five solid (Yin) organs, which are the heart, the lungs, the spleen, the liver and the kidneys.

The Yin organs are responsible for producing, transforming, regulating and storing fundamental substances such as qi, blood and body fluids. Generally, the Yin organs are solid organs without cavities.

Fu is made up of the six hollow (Yang) organs. These are the small intestine, the triple burner (not a single self-contained organ, but a functional energy system), the stomach, the large intestine, the gallbladder (not present in the horse) and the urinary bladder.

The Yang organs exist to digest food, deliver nutrients to the body, and to flush out toxic and waste materials. The Yang organs all have hollow cavities.

Interestingly, the horse has no gallbladder, and bile constantly empties straight into the small intestine even without the presence of food requiring digestion. The horse also continuously secretes gastric acid, which means that it needs to eat constantly. If it is not fed for half a day, it will start to look jaundiced.

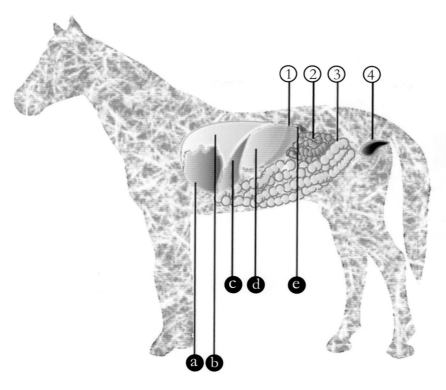

Fu Organs:
1 Stomach
2 Small Intestine
3 Large Intestine
4 Urinary Bladder

Zang Organs:
a Heart
b Lungs
c Liver
d Spleen
e Kidneys

Fig. 5.7 Fu and zang organs of the horse.

Also, the equine kidney and urinary bladder have extremely efficient physiological functions, and renal or bladder diseases in the horse are thus extremely rare.

THE MERIDIAN SYSTEM

The meridian system is a concept in TCM described as an invisible distribution network along which qi flows and through which all body structures – skin, tendons, bone, zangfu, blood, cells and atoms – are connected.

As long as the meridian system runs well, the body (including mind, spirit and emotions) will be healthy and maintain homeostasis where Yin and Yang function harmoniously.

The meridian system in the horse involves 12 principal meridians, each with a symmetrical pair that spread across the chest, back, head, face and extremities (Fig. 5.8).

Despite the absence of a gallbladder, the horse nevertheless has a meridian that corresponds to the gallbladder meridian in humans. This meridian runs through the corner of the eyes, temples, sides of the neck, and then along the body and down the hind legs.

AETIOLOGY

Three basic causes lie at the root of all disease. They are external, internal and miscellaneous (neither internal nor external).

EXTERNAL CAUSES
The external causes are also called external evils and enter the body through the skin, mouth and nostrils to induce superficial disease. Superficial disease is always acute and initially manifests as cold and fever. If it is not dealt with, it will worsen and arouse internal evils.

SIX EVILS
The six evils are those of wind, cold, summer heat, wetness, dryness and fire (heat), which are actually

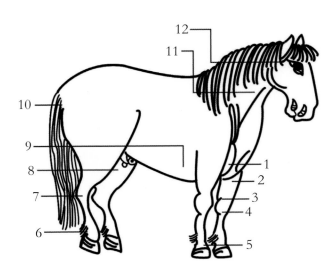

1   Xiong Tang (Thoracic vein) minor Yin heart meridian

2   Tong Jin (Cephalic vein) major Yang small intestine meridian

3   Ye Yan (Chestnut) absolute Yin pericardium meridian

4   Xi Mai (Carpal vein) Yang supreme large intestine meridian

5   Ti Tou (Hoof) minor Yang triple burner meridian

6   Lao Tang (Labour hall) minor Yang gallbladder meridian

7   Qu Chi (Pond on the curve) Yang supreme stomach meridian

8   Shen Tang (Kidney hall) minor Yin kidney meridian

9   Dai Mai (Girdling vessel) major Yin spleen meridian

10 Wei Ben (Tail vein) major Yang urinary bladder meridian

11 Hu Mai (Falcon vessel) major Yin lung meridian

12 Yan Mai (Eye vessel) absolute Yin liver meridian

Fig. 5.8 Twelve principal meridians. (Reference: *Simu's Collection of Equine Medicine*)

natural climatic conditions prevalent in different seasons. Normally, animals can adapt to different conditions and changes of climate without getting sick. However, when the conditions are not as they should be and are either in considerable excess or insufficiency, animals cannot cope and will become ill. Evil implies "excess", i.e. a normal condition develops abnormally to such an extent that the body can no longer tolerate it and it thus turns into a pathogen.

This does not mean that an abnormal condition always gives rise to a disease; it is only those animals without sufficient healthy qi that will be affected. Even under normal climatic conditions, an animal without enough healthy qi, a healthy body or proper adaptability, will also become sick.

Of the six evils, wetness is the most threatening evil to the horse. Wetness is Yin in nature, sinking, turbid, heavy, sluggish and sticky.

As summer wanes, the period from the start of autumn to the autumnal equinox is known as late summer. At this time in China, the humidity is highest and wetness evil is widespread. On top of this, if a horse gets soaked by rain or stays in wetland areas for a long time, the chances are that it will catch wetness evil.

Wetness evil will produce different pathological changes. If the evil lodges in a meridian, the horse will have heavy limbs, swollen and painful joints, and difficulty in stretching and moving. When the evil lodges in the spleen and stomach, the horse will have a poor appetite, loose stools, a heavy and sluggish body, as well as display pale-yellowish gums. Its skin will also likely appear eczematous. Diseases caused by the wetness evil are chronic. They are difficult to treat and often relapse, as is commonly the case in eczema and arthralgia.

Summer heat is the second-most threatening evil to the horse. Summer heat occurs between the summer solstice and the onset of autumn. It is Yang in nature, characterized by scorching sunshine, humidity, ascension and dispersal.

When summer heat attacks an animal, it will produce excessive perspiration, thirst and a strong and rapid pulse.

Upper parts of the body are susceptible to the attack as summer heat tends to rise and disperse, striking the head and eyes, resulting in irritation and anxiety.

Summer is often humid and rainy in China, and summer heat evil is frequently compounded by wetness evil. Typical signs of summer heat evil include fever, irritation and thirst, loss of appetite, loose stools and fatigue.

Jaundice is a disease often seen in the family *Equidae* and exhibits as a yellowish colour in the whites of the eyes and in the gums. It is caused by an attack of summer heat and wetness. This disturbs the spleen and stomach, and then the liver. If the liver then fails to regulate its bile production, bile will spill over into the bloodstream, bringing on jaundice.

Gastroenteritis is also a common disease seen in the family *Equidae*. When summer heat lodges in the gastrointestinal tract, fever, diarrhoea and abdominal pain are likely to develop.

EPIDEMICS

Epidemics result from infectious or contagious diseases which are transmitted to horses through the air and by contact with infected food, blood and insect vectors, such as blood-sucking horse flies, which can cause viral anaemia.

Once an outbreak takes place, the disease can spread quickly. Every horse is susceptible irrespective of its age, gender and physical condition. Epidemics are acute, malicious and mutable. Anthrax is a typical epidemic seen in horses. Clinical changes include fever, colic, and a swollen neck, throat and stomach. This often-fatal disease soon progresses to cause rapid breathing and coma. The horse will invariably die if immediate veterinary attention is not provided.

Extreme weather, especially severe and unseasonable episodes like droughts, intense heatwaves, flooding and persistent fog can lead to the outbreak of epidemics. Pollution is another causative factor. Exposure to contaminated water and food or polluted air also plays a part. Additionally, inadequate immunization and preventive measures all render epidemics prevalent or potentially emergent.

INTERNAL CAUSES

The internal causes mainly pertain to husbandry problems and can result in healthy qi deficiency, weakening immunity and increasing vulnerability to external evils.

DIET

Starvation, inappropriate feeds, malnutrition and imbalanced diets all compromise the health of a horse. If incorrectly fed, the animal will end up with underdeveloped organs, reduced productivity and stunted growth. Severe cases can lead to the animal's collapse or even death.

Overfeeding is another problem. Horses have a relatively small stomach for their size compared with other livestock such as pigs and cattle. Physically unable to vomit or belch, horses must be fed relatively small amounts at frequent intervals. Overeating will cause dysfunction of the stomach and intestines.

EXERCISE AND LABOUR

Excessive labour or exercise with inadequate rest can sap a horse's qi and blood, affect the function of the spleen and stomach, and finally give rise to mental and physical exhaustion. Idleness is equally disabling. When living wild, horses naturally walk, trot and occasionally gallop at whim. They may have to travel 20 to 30 kilometres every day in the course of just getting enough to eat and drink. However, horses that are confined in a stable do not have the same space to exercise themselves. Regular exercise or labour is then essential for all aspects of their general fitness. From a TCVM point of view, idleness will result in stagnation of blood and qi, a decline in the function of visceral organs, accumulation of fat and eventually impairment of productivity and immunity. A lethargic stallion's sperm is less active and thus its fertility is lower. An overweight mare is more likely to be infertile or to suffer a miscarriage.

OTHER CAUSES

BITES AND INJURIES

Fights between horses, bites in the courtship ritual, abrasions from fencing and leg injuries sustained during exercise are common superficial horse injuries. In addition, bites from rabid animals, bees, snakes and scorpions are possible, although horsefly bites are the most common.

Poisoning in horses is not a common problem as most horses tend to avoid toxic plants unless they are really hungry. *Achnatherum inebrians* (Hance) Keng, commonly referred to as drunken horse grass, is a highly drought-tolerant noxious weed of western China. An amount consumed by a horse that is more than 1 per cent of its own weight is poisonous and may be fatal. Droughts and pasture degradation alike can enhance the growth of drunken horse grass.

Human error can also be a cause of poisoning. Overdosage of salt, additives, cottonseed cake or meal supplements, or feeds with impurities or unintended substances all pose a toxic risk to horses kept in stables.

PHLEGM

Called in Chinese *tan yin* (tan meaning viscous and turbid and yin thin and clear), phlegm is caused by accumulation and transformation of body fluids. It can be a product of visceral organ imbalance and may also be a cause of disease.

Other than dysfunction of the lungs, spleen and kidneys, which are involved in the distribution and discharge of water, the fire evil's alteration of body fluid composition, stagnation of body qi and congestion of meridians can all give rise to phlegm.

If a horse gasps and coughs, then the chances are that phlegm has accumulated in its lungs. If the animal has water retention, phlegm will have accumulated in the skin. Diarrhoea suggests phlegm is present in the intestines.

BLOOD STASIS

Blood stasis is when blood congests rather than flows or it accumulates in areas of the body and is slow to circulate. Like phlegm, this can be a result of disease and also a cause of disease. Yang qi deficiency, sluggish qi, hot blood with reduced body fluids, and meridians attacked by the cold evil can all cause stagnation of blood flow. Blood stasis will manifest as swelling, pain and skin discolouration.

Ringworm, fleas, mites, lice, bot flies and ticks are common external horse parasites. They are vectors of disease and can all cause a myriad of harms, which include anaemia, weight loss, irritation, scratching, dermatitis and digestive deficiency. Wet, dirty and damp conditions encourage the breeding of parasites.

All horses harbour internal parasites. According to the University of Arkansas, horses host the largest worm burden at 100,000 organisms. Ninety per cent of all colic cases may be associated with rupture or circulation blockages caused by the migrating larvae of *Strongylus vulgaris* (blood worms). Fifty per cent of deaths in horses may be associated with internal parasites. Most internal parasites are host-specific. Horse parasites can only complete their life cycle in horses but not in cattle or dogs. If a horse parasite enters a cow's body, the organism's life cycle is disrupted and it will die.

Large and small strongyles, ascarids, bots and pinworms are the five major equine internal parasites and the large strongyles (blood worms) are the most common and most destructive.

FOUR PRINCIPAL METHODS OF DIAGNOSIS

There are four principal methods of diagnosis in TCVM. These are inspection, auscultation and olfaction, interrogation, and palpation and pulse taking.

The methods are used in combination as a comprehensive system to diagnose by identifying clinical signs. Each method helps gather different signs and information. Neglect of any one method can render valueless a diagnosis based on the other methods and lead to an overall misdiagnosis.

INSPECTION

First, clinicians should observe the animal from a distance of about 1.5 to 2 metres. They should evaluate the animal as a whole, and then its spirit, body, coat, posture, breath, chest and stomach.

After that, they assess specific parts of the body from front to back and left to right.

WHOLE BODY INSPECTION

▫ Spirit. The spirit is the manifestation of an animal's vitality. It is observed mainly through the eyes and ears. The horse's spirit can reflect visceral organ conditions and disease severity. When a horse is in good spirits, its eyes are open and bright. The inside of the eye is pink; the ears are alert; and the reaction is agile and quick. If an ailing horse's healthy qi is impaired and manifests as low spirits, cloudy eyes, drooping ears and head, as well as a sluggish reaction, it is an indication that the disease is worsening.

▫ Physical condition and posture. The healthy horse has a smooth circulation of qi and blood, a shiny, soft and smooth coat, moist and resilient skin and agile limbs.

The healthy horse stands on all fours and rests a hind foot when it is relaxed or bored. It lies down occasionally, but when approached, it stands up straightaway. Once it has been untacked, it loves to roll, rise and then shake off the excess dirt.

Repeated lying down and rising, pacing, rolling, pawing, scraping and turning of the head to look at the stomach and hindquarters are clinical signs of colic (Fig. 5.9). Exhibiting a short and stiff gait with rapid foot placement, standing in a "founder stance", i.e. shifting weight to the other legs, with the affected leg or legs extended, reluctance to move and a tendency to lie down are signs of laminitis.

▫ Superficies (skin and coat). As the lungs are associated with the outward appearance of the horse, complexion, lustre, texture, moisture of the skin and coat are indicative of the condition of qi, blood and lungs.

The healthy horse has a resilient and moist skin, and a glossy and soft coat that lies flat and smooth, growing and shedding in keeping with the seasons of the year.

Itchy skin and wheals containing a yellowish watery secretion are a sign of urticaria. This results from weakened lung alveoli and qi failing to diffuse. A contracted skin and raised hairs suggest the lungs are being constrained by an exogenous cold evil. A withered skin, a dull and coarse coat, and a thick winter coat unshed in summer are all signs of deficient qi and blood, and malnutrition.

INSPECTION OF ANATOMICAL PARTS

▫ Eyes. As indicated above, the healthy horse has clear and bright eyes, with no discharge, and the inside of the eye is pink. When the eyes are swollen and reddish, sensitive to light, watery and difficult to open because of an abundant yellowish discharge, there is too much liver heat.

Swollen eyelids are caused by water retention. Reluctance to open the eyelids, a drooping head and floppy ears are signs of overexertion, and chronic and serious diseases. Sunken eyes are caused by depletion of body fluids. Pupil dilation can be an indication of a life-threatening condition such as total exhaustion of qi and essence, and poisoning.

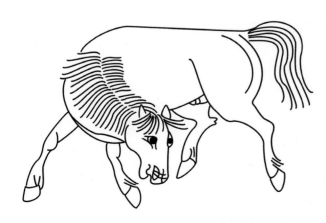

Fig. 5.9 A clinical sign of colic: turning the head towards the stomach and hindquarters. (Reference: *Simu's Collection of Equine Medicine*)

□ Nostrils. The nostrils are the outlet of the lungs. The healthy horse has clean and moist nostrils, breathing smoothly and silently.

A thin, clear nasal discharge signals an attack of wind and cold. A thick and sticky nasal discharge, on the other hand, signals an attack of wind and heat.

The passing of a persistent, yellowish-white and turbid discharge with a fishy smell in one nostril is a sign of frontal sinusitis. If both nostrils produce a profuse, thick discharge and swollen lymph nodes occur in the lower jaw, especially with an abscess, this suggests the animal has strangles (equine distemper).

□ Ears. The healthy horse has a pair of alert ears and a normal sense of hearing. Droopy ears indicate a long-term critical illness or frail kidney qi. A horse that looks scared and anxious while its ears are pricked upright may have an attack of heat evil or tetanus. Visible blood vessels behind the ears through to the tips are indicative of an exogenous hot syndrome. Ears that are cold and have no visible blood vessels behind them suggest an exogenous cold syndrome. Floppy ears and asymmetrical eyes and mouth indicate unilateral facial paralysis.

□ Mouth and lips. The mouth and lips form a corresponding pair with the spleen. The healthy horse has a moderate amount of foam and saliva in the mouth, implying its acceptance of the bit. However, an excessive flow of saliva, dribbling of saliva and reduced saliva are all signs that something is wrong.

A slack lower lip signals spleen deficiency. Crooked lips are associated with stroke. Oral ulcers or lesions indicate exuberant heat in the heart meridian.

□ Diet. The horse's appetite can reflect the condition of its stomach. With normal stomach qi, a healthy horse has a relentless appetite. If a sick horse maintains a normal appetite and stomach qi, the animal will have a positive prognosis, and vice versa.

If a horse eats only grass, it may develop a sore mouth, or even renal failure. If a horse favours only a dry feed, it suggests a cold and wet syndrome in the stomach and spleen. If a horse favours a wet feed, there is heat present in the stomach. If a horse chews slowly, spits while eating and swallows with difficulty, it has a sore throat or dental problems. If a horse eats sand, soil, hair, faeces and other such materials, it suggests that its diet lacks minerals or trace elements.

□ Breathing. The normal breathing rate of a healthy horse at rest is eight to 16 breaths per minute. Breathing is directed by the lungs and associated with kidney qi. Frequency, intensity, rhythm and posture in breathing should all be observed. Slow and shallow breathing and shortness of breath after a little exercise indicate a deficient or cold disorder. Heavy, rapid and laboured breathing suggests an excessive heat syndrome. Obvious stomach movement during breathing is indicative of chest pain, while obvious chest movement is indicative of stomach pain. When an ailing horse has a long inspiration and short expiration, its qi and blood are connected and suggests its natural qi is still sufficient to enable it to recover even from a critical condition. Dyspnoea indicates lung failure and a critical condition.

□ Faeces. The healthy horse defecates 10 to 12 times a day, producing between 15 and 23 kg of faeces. If the horse is fed on green roughage, the faeces will be greenish; if its feed consists of light-coloured hay, its faeces will be more yellow. Healthy faeces are ball shaped and will partially crumble when they drop to the ground.

The quantity, colour, smell and shape of faeces in general should be consistent; any changes may suggest stomach and intestinal disorders.

□ Urine. Examination of the horse's urine involves observation of the quantity, colour and turbidity. Horse urine is normally yellowish and is often turbid due to the presence of large amounts of calcium carbonate. High turbidity actually reflects problems in the kidneys, anus or reproductive organs.

□ External genitalia and anus. The external genitalia or reproductive organs include the stallion's penis and testicles and the mare's vulva.

A hard and swollen scrotum and testicles are common signs of Yin kidney inflammation, which is often seen in horses in autumn and winter.

If the scrotum feels very hot and is painful to the horse on touch, the horse may have Yang kidney inflammation. Premature ejaculation is indicative of kidney qi deficiency. Erectile dysfunction can be due to a deficiency of the liver and kidneys. Paraphimosis (inability to retract the penis) is related to a cold and deficient kidney meridian.

The vulva should be examined for shape, colour and secretions. During oestrus, the vulva will turn red and swell, releasing a small amount of secretion. If lochia persists long after the postpartum period, immediate veterinary attention is needed. During gestation, a swollen and elongated vulva producing a yellow or white discharge is a sign of miscarriage. When one side of the vagina is sunken and the animal has abdominal pain, the uterus is most likely to have become twisted.

Examination of the anus focuses on the organ's tightness, flexibility and adjacent areas. If the anus is loose and sunken, this suggests qi deficiency and chronic diarrhoea. An itchy anus and tail rubbing can be a sign of pinworm infestation. Faecal soiling of the anal area, tail root and hocks is indicative of diarrhoea.

ORAL CAVITY

Examination of the oral cavity is an important feature of TCVM and is especially helpful in the diagnosis and prognosis of equine diseases. This includes inspection of the lips, gums, tongue, sublingual caruncles and tongue coating (Fig. 5.10).

□ Colour. The healthy horse has a pink tongue. The gums are slightly paler than those of humans and pigs, but are darker than those of cows, camels and sheep. The colour varies with season and age.

In general, the gums are paler in winter and in old horses, but darker in summer and in young horses.

The most commonly seen abnormal colourations range from white and greyish-white to bright yellow, reddish-brown and black. Each signifies a different disharmonic state.

□ Coating. The colour and texture of the tongue coating are the two main areas of examination interest in the animal's mouth.

The healthy horse has a thin white or pale-yellowish tongue coating. Unhealthy colours are heavy white, yellow, grey and black.

The absence of a coating, or one that is too thick or too thin, too moist or too dry, smelly or greasy, is suggestive of an unhealthy animal.

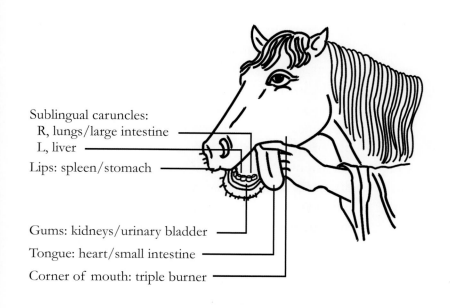

Sublingual caruncles:
  R, lungs/large intestine
  L, liver
Lips: spleen/stomach
Gums: kidneys/urinary bladder
Tongue: heart/small intestine
Corner of mouth: triple burner

Fig. 5.10 Inspection of the oral cavity gives clues as to the affected organ. (Reference: *Yuan Heng's Collection of Horse Treatments*)

□ Fluids. The presence of saliva reflects moistness of the mouth cavity and the body's fluid level.

If a horse's saliva is viscous or insufficient, it is indicative of heat penetrating Yin. Excessive, thin, clear saliva accompanied by coolness in the mouth, signifies a cold syndrome or water retention. If the mouth cavity is excessively moist, slippery, viscous and hot, there is profuse internal damp-heat. Excess saliva results from Yang deficiency, too much dampness in the spleen and stomach, or oral disease.

□ Tongue. A healthy tongue is slightly moist, soft and flexible. It should be neither fat nor lean, and not slippery or dry, but should have a thin white coating.

If the tongue is pale and fleshy, with teeth marks along the rim, it is suggestive of Yang deficiency in the spleen and kidneys. A red and swollen tongue with ulceration is a sign of heart fire uprising, while a thin, tender and whitish tongue with a thin coating is a result of deficient qi and blood. If the surface of the tongue is cracked, it is an indication of hot disease. A hard and rigid tongue is caused by activity in hot evil qi, which penetrates the pericardium and depletes Yin blood, whereas a shrivelled tongue suggests liver disease or a chronic disease resulting from deficient qi and blood. If the tongue is very pale, atrophic and cannot be put out or fails to suck back saliva, the horse is in a critical condition with deficient qi and blood.

AUSCULTATION AND OLFACTION
Auscultation relies on listening to the sounds that a horse makes. Olfaction is used to check the smell of the animal's secretions and discharges.

AUSCULTATION
□ Sounds. If a horse wheezes in the early stages of a disease, it suggests the disease is caused by an exogenous wind-cold evil. Loss of sound in the wake of a chronic disease implies deficient lung qi. A faint low sound is also a sign of deficiency. However, a coarse and deep turbid sound signals the result of an excess of qi. A sound which starts flat and is prolonged at the end is suggestive of

a positive prognosis. Strange but short sounds indicate an attack of heat evil to the heart and a negative prognosis. If a horse is groaning, chewing air or grinding its teeth, it is invariably in pain or is in a critical condition.

□ Breathing. If the breathing is feeble, it is a clinical sign of internal injury, deficiency and exhaustion. A long inhalation followed by a short exhalation is indicative of the presence of healthy qi. A short inhalation followed by a long exhalation indicates deficient healthy qi, and kidney and lung failure. When breathing comes with the sound of a stuffy nose, it suggests swelling or a lesion in the nasal passage or excessive rhinorrhoea (discharge of watery mucus from the nose). If a horse has an audible phlegm rale, it is a sign of phlegm stagnation. A fishy-smelling nasal discharge together with heavy and deep breathing, or an open mouth and broad nares alongside see-saw breathing, are indicative of a critical condition with a negative prognosis.

□ Coughing. Noisy coughing suggests congestive or febrile disease, occurring more often in an early wind-cold or wind-heat syndrome caused by exogenous factors. A faint low cough is a sign of deficiency and more often occurs in excessive labour and chronic coughs. A productive cough producing phlegm is indicative of cold in the lungs or tuberculosis. A dry cough without phlegm often happens in Yin deficiency when the lungs are dry or in the early stages of heat in the lungs. A persistent, faint and feeble cough with a thick nasal discharge and see-saw breathing suggests serious illness. Coughing is always a clinical sign of lung dysfunction.

□ Intestines. The intestinal sounds of the horse are clearly audible. In enteritis, there is an increase in the intensity and frequency of the sounds and they have a distinct fluid quality. A horse suffering from spasmodic colic produces a loud and almost crackling sound. A decrease in the intensity and frequency of borborygmus (rumbling) occurs in impaction colic of the animal's large intestine.

Complete absence of abdominal sounds is possible in thromboembolic colic due to a verminous aneurysm (inflammation caused by worms) and infarction (obstruction of the blood supply) of the large intestine.

OLFACTION

▫ Breath. The breath of the healthy horse has a grassy smell. Bad breath, a hot mouth and anorexia indicate excessive heat in the stomach and intestines. A sour smell suggests a stagnant stomach. Fishy and rancid smells occur with mouth lesions, gum diseases and abscesses of teeth roots or gums, and need further investigation.

▫ Nostrils. The healthy horse does not produce any particular smell from the nostrils. A yellowish, thick and fetid nasal discharge suggests heat in the lungs, while a yellowish-grey and fishy nasal discharge accompanies a pulmonary abscess. A greyish-white, cottage cheese-like discharge that smells like rotting flesh occurs in pulmonary failure.

If there is a fetid and heavy discharge in only one nasal passage, it suggests that the paranasal sinuses contain thick pus.

▫ Faeces and urine. Fluid and slightly foul faeces indicate a deficient and cold syndrome. Rough and sour-smelling faeces often follow indigestion caused by overeating or an improper diet. Production of paste-like and fetid faeces flecked with blood or mucus is a sign of damp-heat.

The urine of the horse is ammoniacal. Long and clear streams of urine lacking any smell of ammonia occur in a deficient or cold syndrome. Small volumes of turbid and reddish urine with a pronounced ammoniacal odour are produced in an excessive or hot one.

▫ Pus. Benign lesions discharge pus that is yellowish-white, bright, odour free or slightly foul. In an excessive and Yang syndrome where internal fire-poison is active, the pus is yellowish, viscid, turbid and fetid. In a deficient and Yin syndrome where pathogenic evils remain and qi and blood weaken, pus is ashen, thin and fishy.

INTERROGATION

Interrogation of stable managers, grooms, owners and riders can yield helpful information for diagnosis.

Clinical signs, disease progress and treatments received are the focus of enquiry. Diet, faeces, urine, coughing, pain, gait, cold intolerance, fever, perspiration or inability to sweat all offer useful information. A good appetite implies less severity; loss of appetite is usually a bad sign. Coughing and gasping which are noticeably worse in the evening than in the morning occur in cases of deficient Yang. Conversely, if they are worse in the morning than in the evening, this is a sign of excessive Yang. Chronic diseases, unstable appetite, faeces which sometimes turn dry while at other times are soft and weight loss are all signs of a frail spleen and stomach.

Attention to medical records is very important as immunity can result from previous infection or vaccination.

Likewise, breeding records, husbandry and workload all provide clues for a successful diagnosis.

Pulse Taking and Palpation

Pulse taking and palpation are done by touching the horse with the fingers and palm and noting the tactile signals.

PULSE TAKING

The horse's pulse can be best felt near the front of the left jawbone. Not only is the rhythm of the pulse measured but also its depth, speed, strength, slipperiness and intensity. Combined, these indicate the nature, characteristics and location of any disease. There are two types of pulse pattern: normal and pathological.

▫ Normal pulse pattern. The pulse pattern of a healthy horse is relaxed, harmonious and even. It should be neither rapid nor slow, and not too strong or weak.

The normal pulse pattern varies with the seasons. It is said that the spring pulse is tense like a bowstring; the summer pulse surges like a hook;

the autumn pulse floats like a hair; and the winter pulse sinks like a stone.

The pattern is also influenced by factors such as age, gender, size, work life and body fat. A foal has a rapid pulse. An old or frail horse has a weaker pulse. A thin horse tends to have a floating pulse while a fat horse is more likely to have a sinking pulse. The pulse rate tends to increase after exercise, to weaken during starvation, to surge after feeding and to become smooth during pregnancy.

The average resting pulse rate of an adult horse (aged four to 20 years) is 30 to 40 beats per minute (bpm) depending on the breed. The average pulse rate for a foal is 70 to 120 bpm, and for a yearling, it is 45 to 50 bpm.

A horse's heartbeat rises to 180 to 240 bmp during or immediately after exertion and gradually returns to normal within 10 to 20 minutes.

□　Pathological pulse pattern. There are six basic pathological pulse patterns, which are grouped into three opposite pairs. Other than featuring individually, each pattern can at times develop in combination with one or more patterns from the other pairs.

Floating and Sinking. A floating pulse can be easily detected by touching the skin lightly, but the tactile reflex is lost by pressing hard. It is seen in an exogenous syndrome caused by the six evils that enter the body through the skin, mouth and nostrils. A sinking pulse can be detected only by pressing hard on the artery. It is felt in an endogenous syndrome caused by a direct attack of the six evils on the visceral organs, or provoked by such internal evils as starvation, overfeeding and excessive labour.

Slow and Rapid. In an adult horse, a slow pulse is one that is less than 30 bpm; a rapid pulse is one higher than 45 bpm. A slow pulse occurs in a cold syndrome while a rapid pulse occurs in a hot one.

Empty and Solid. An empty pulse is sensed when the blood vessel seems flaccid, suggesting a deficient syndrome like blood and qi deficiency. A solid pulse is sensed when the blood vessel is full and forceful, suggesting an excessive syndrome like fever, constipation, and qi and blood stagnation.

PALPATION

Palpation uses the sense of touch through the palm to check the animal's temperature and suppleness and to detect any pain or anatomical abnormalities.

□　Temperature. The mouth cavity, nose, ears, superficies and limbs are examined with bare hands to determine whether a hot, cold, excessive or deficient syndrome is present.

□　Distension. Any distension should be defined by its nature, size, shape and sensitivity. A hard distension can mean an osteoma (bony outgrowth). A soft and elastic distention may be a blood tumour or an abscess. Fluid retention feels doughy, and applying pressure leaves an indentation at the pressure point which is slow to recover. Emphysema is indicated by the soft texture of a swelling in the lung region and crepitation (rattling sound). A swelling which is acutely painful, prominent or feels hot relates to a Yang syndrome, while a swelling which is contracted or slightly painful is that of a Yin syndrome.

□　Thorax and abdomen. If a horse is sensitive, backs away or coughs during percussion of the thorax, it has problems in the lungs or thoracic wall, such as a pulmonary abscess, or chest or diaphragm pain. If it is only sensitive on one side and has no cough, it is probable that the thoracic wall is injured.

An abdominal swelling that when struck produces a drum-like sound indicates gas in the abdomen. Abdominal fullness and firmness signal food stagnation in the stomach and spleen. Rigidity, pain and a fluid, splashing sound on palpation of the abdominal wall are typical signs of peritonitis. A swollen breast that is hard and hot suggests acute mastitis.

□　Rectal palpation. In rectal palpation of the horse (as with the cow), the hand and arm are inserted into the rectum to examine the reproductive tract, gastrointestinal tract, kidneys, pelvis or lumbo-sacral spine.

It is one of the most useful diagnostic procedures in a horse with obstructive colic as it can provide information as to the cause and suggest treatment guidelines. Rectal exploration is also used to break up manually or remove impacted faeces lodged in the lower gastrointestinal tract.

METHODS OF TREATMENT

Traditional Chinese veterinary medicine is used both internally and externally to treat disease. Internal treatment is primarily given with the horse's ingestion of herbal mixtures, many of the ingredients of which are also found in Western pharmacopoeias. External treatment includes acupuncture, massage, cauterization, bloodletting, qigong and the application of herbal remedies. Each method can be effective on its own or can be used in combination with any of the other methods.

In China, herbalism and acupuncture are the most commonly used TCVM treatments.

HERBALISM

Horses are herbivores. In a perfect, natural environment like the steppes of Mongolia, the prairies of North America and the moors of the Shetland Isles, where horses can roam over a large area and obtain a diverse variety of grass and plants, they can instinctively self-medicate by selecting and eating what plants they need to prevent and treat disease. However, the natural environment increasingly is becoming degraded and the average domestic horse never has the chance to practise its own inbred understanding of plant lore.

Herbal medicine is in keeping with the nature of horses, whether applied internally or externally. Internal methods mainly involve the ingestion of herbs as powders, decoctions or infusions. External methods include the application of herbal powders and soft plasters on the horse's body, herbal massages, washes and compresses.

The active ingredients of Chinese herbal medicines are found in the different parts of plants – the leaf, stem, flower, fruit and root – depending on the species (Fig. 5.11). The plant material is always supplied dried. Ingredients are also obtained from animal and mineral sources. Herbalism is generally formula based with many herbs and other ingredients tailored to an individual patient.

The theory of TCVM holds that every medicinal substance has a positive and negative pharmacological effect. Ingredients in a prescription should be compatible with each other in order to accentuate efficacy while reducing side effects.

ACUPUNCTURE

The concept of acupuncture is based on maintaining health with a balanced flow of qi. Through stimulation of specific acupuncture points on the horse's body by insertion of sterile, hair-thin needles along meridians, the flow of energy is restored, achieving a therapeutic effect. Among some 360 acupuncture points located throughout the equine body, about 150 are commonly used and most of them will be found on meridians (Fig. 5.8).

Acupuncture can help a wide range of diseases and health conditions. It is especially beneficial to horses with colic, lameness and musculoskeletal pain and is gaining in acceptance among Western veterinarians.

OTHER THERAPIES

MASSAGE

Based on the same concept as acupuncture to maintain a balanced flow of qi, massage therapy employs fingertip pressure rather than needles. Depending on the ailment, different forms of pressure are applied to stimulate particular meridians or acupoints. When practising massage, short regular sessions rather than infrequent longer sessions are advisable.

Massage for horses is commonly used to enhance performance levels and endurance, prevent injuries and speed up recovery.

BLOODLETTING

Bloodletting from a vessel on the neck in the spring, taking 1000 to 2000 ml of blood, was said to help prevent disease in the year ahead. However, most owners nowadays view the practice as archaic and cruel, and it is thus seldom performed. The technique was also used to treat heat stroke, acute heart failure and pulmonary haemorrhage, among other conditions.

Lotus Plumule

Arctium Fruit

Chinese Raspberry

Chrysanthemum

Bitter Orange

Loquat Leaf

Sappan Wood

Vitex Fruit

Pagoda Tree Seed

Mulberry Mistletoe Stem

Tangerine Peel

Evodia Fruit

Lotus Root

Gentian Root

Motherwort

Honey Locust Thorns

Solomon's Seal

Knotgrass Stem

Chinese Gentian Root

Trichosanthes Peel

Milletia Vine Stem

Quince Fruit

Lily Bulb

Madder Root

Fig. 5.11 Herbs often used in TCVM, common English names.

COMMON EQUINE DISEASES: DIAGNOSIS AND TREATMENT

As horse racing and other Western equestrian pursuits such as showjumping, cross-country events and dressage are less common in China, leg injuries, for example bowed tendons, overreaching (when the toe of the hind foot extends forward and strikes the forefoot on the same side), abrasions to the hind legs and inflamed forelegs, occur infrequently. More commonly encountered in draught horses and club horses in China are spasmodic colic, obstructive colic and gastroenteritis.

SPASMODIC COLIC

"Colic" is a general term that refers to any disease process that gives rise to abdominal pain. Among the different types of colic, spasmodic and obstructive colic are the most commonly seen in the family *Equidae*.

Fig. 5.12 A sign of spasmodic colic: turning to look back at the stomach. (Reference: *Simu's Collection of Equine Medicine*)

CLINICAL SIGNS

The signs of spasmodic colic include severe abdominal pain produced by contraction, or spasm, of a portion of the small intestine.

The bouts of pain come and go, often with a few hours of respite between bouts. During the intervals of ease, the horse can appear entirely normal, but before long the pain returns and gets worse, the intervals of ease becoming shorter and shorter.

Loud, almost crackling intestinal sounds can be heard if the ear is placed over the horse's flank. The horse will be pawing at its belly with a hind hoof, and will suddenly lie down, roll and rise up. It may also turn to look back at its stomach (Fig. 5.12) and shiver all over. The ears and nostrils will feel cold, the oral cavity will be pale, and the pulse will be sinking and slow.

AETIOLOGY AND PATHOGENESIS

Spasmodic colic can be caused by an attack of exogenous cold and wind pathogens, such as a sudden drop in temperature or a bad soaking in the rain. Internal injury to the spleen and stomach due to Yin cold brought on by drinking too much chilled water, eating too much chilled forage or a soaring body temperature after overexertion can also cause spasmodic colic.

TREATMENT

Treatment involves warming the central part of the horse's abdomen — the spleen, stomach and liver — to dispel the cold evil and regulate the qi and blood flow.

▫ Herbal treatment. Ju Pi (Tangerine Peel) Powder

Ingredients and action:
Citrus reticulata (Qing Pi, green peel) 25 g
    Stimulates qi flow
Citrus reticulata (Chen Pi, ripe peel) 30 g
    Stimulates qi flow
Magnolia biondii (Hou Po) 30 g
    Promotes qi flow and warms body centre
Cinnamomum cassia (Gui Xin) 15 g
    Warms body centre
Asarum heterotropoides (Xi Xin) 5 g
    Dispels cold and relieves pain

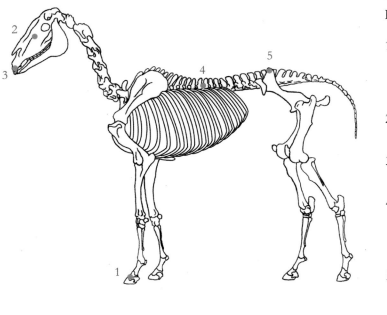

Region	Acupuncture Point/Action
1 Hoof	Ti Tou (Hoof): Relieves abdominal pain, bloat or accumulation of gas in the stomach, indigestion and depressed appetite
2 Head	San Jiang (Three rivers): Relieves abdominal pain and distension
3 Upper Lip	Fen Shui (Dividing water): Relieves spasmodic colic
4 Body	Pi Shu (Spleen transporter): Relieves constipation, abdominal distension, food stagnation, diarrhoea, indigestion and depressed appetite
5 Body	Bai Hui (Hundred convergences): Relieves constipation, abdominal distension, diarrhoea and spasmodic pain

Fig. 5.13 Acupuncture points used for treating spasmodic colic.

Foeniculum vulgare (Hui Xiang) 30 g
 Dispels cold and relieves pain
Angelica acutiloba (Dang Gui) 25 g
 Invigorates blood and smoothes qi flow
Angelica dahurica (Bai Zhi) 15 g
 Dispels cold and relieves pain
Areca catechu (Bing Lang) 15 g
 Promotes qi flow and resolves stagnation

The ingredients are ground together well and given mixed with spring onion, salt and vinegar, which act as catalysts to promote herbal action.

The prescription is based on a classical recipe from *Yuan Heng's Collection of Horse Treatments*.

▫ Acupuncture. Several acupuncture points (Fig. 5.13) are used to treat spasmodic colic in the horse. The chart above describes their names and mode of action.

▫ Supplementary measures. Repeated lying down and rising up as well as rolling can cause the horse to stumble and hurt itself or result in displacement of the stomach (torsion). It is important to get a colicky horse on its feet to prevent injuries.

Walking the horse to stimulate digestion and blood circulation can also help.

To prevent a horse becoming colicky, it is important not to let it drink too much chilled water immediately after exertion and to avoid it resting in wet, cold and windy places.

OBSTRUCTIVE COLIC

CLINICAL SIGNS

In obstructive (also known as impaction) colic, partially digested food or dry, firm faecal material accumulates in the intestines, preventing the normal passage of gut contents and the disruption of qi movement. Signs include anxiety, loss of appetite, small and dry faecal balls covered with mucus, abdominal pain, repeated lying down and rising, pacing, rolling (Fig. 5.14), pawing and/or scraping, turning of the head to view the stomach and hindquarters, hind legs kicking the abdomen, diminished or absence of intestinal sounds (as the intestines move less or cease to move entirely), a dry mouth, stagnation of faeces, a thick and yellowish tongue coating, and a sinking and excessive pulse.

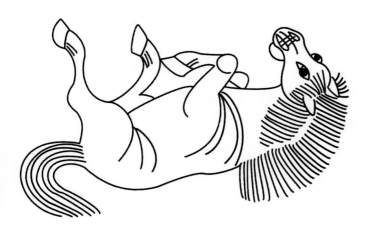

Fig. 5.14 A sign of obstructive colic: rolling on the ground. Seen also in spasmodic colic. (Reference: *Simu's Collection of Equine Medicine*)

AETIOLOGY AND PATHOGENESIS

Obstructive colic is brought about by an improper diet or poor husbandry, body fluids stricken by heat pathogens, penetration of cold pathogens, and blood and qi deficiency.

TREATMENT FOR SMALL INTESTINAL IMPACTION

The treatment of small intestinal impaction involves regulation of qi movement, transformation of undigested food and nourishment of the small intestine.

▫ Herbal treatment. San Xiao Cheng Qi (Qi-promoting Three-digestant) Decoction

Ingredients and action:
Crataegus laevigata (Shan Zha, ground) 60 g
    Removes food stagnation
Hordeum vulgare (Mai Ya, ground) 60 g
    Removes food stagnation
Massa fermentata medicinalis (Liu Qu, ground) 60 g
    Relieves stagnation and promotes qi flow
Rheum officinale (Da Huang) 60 g
    Purges fire, clears heat, invigorates blood

Citrus aurantium (Zhi Shi, immature fruit) 15 g
    Promotes qi and resolves food stasis
Magnolia biondii (Hou Po) 15 g
    Relieves stagnation and promotes qi flow
Areca catechu (Bing Lang) 10 g
    Relieves stagnation and promotes qi flow
Haematitum (Dai Zhe Shi) 45 g
    Brings down unwanted rising qi

The decoction is made by simmering the ingredients in water. The strained liquid is given to the horse.

TREATMENT FOR PELVIC FLEXURE AND CAECUM IMPACTION AND LARGE INTESTINE OBSTRUCTION

The treatment of pelvic flexure and caecum impaction and large intestine obstruction involves the promotion of bowel evacuation.

▫ Herbal treatment. Zhi Shi Po Jie (*Citrus aurantium* Obstruction Removal) Powder

Ingredients and action:
Citrus aurantium (Zhi Shi) 60 g
    Promotes qi and resolves food stasis
Rheum officinale (Da Huang) 30 g
    Purges fire, clears heat, invigorates blood
Cassia angustifolia Vahl (Fan Xie Ye) 30 g
    Drains heat and eliminates stagnation
Pharbitis purpurea (Er Chou) 30 g
    Eliminates excess fluids
Magnolia biondii (Hou Po) 30 g
    Relieves stagnation and promotes qi flow
Citrus reticulata (Qing Pi) 30 g
    Relieves stasis and breaks up lumps
Saussurea lappa (Mu Xiang) 30 g
    Promotes qi to relieve pain
Natrii sulfas (Mang Xiao) 150 g
    Moistens and softens any hard mass, clears heat

The ingredients are ground together into a fine powder and given mixed with water.

TREATMENT FOR RECTAL OBSTRUCTION

The treatment of rectal obstruction involves the relief of dryness, promotion of smooth bowel movements, purgation of the large intestine to resolve food stasis and regulation of qi movement to relieve pain.

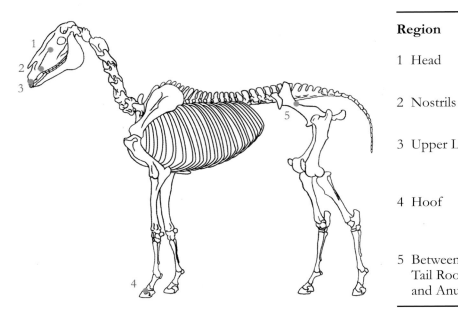

Region	Acupuncture Point/Action
1 Head	San Jiang (Three rivers): Relieves abdominal pain and distension
2 Nostrils	Jiang Ya (Ginger bud): Relieves abdominal pain
3 Upper Lip	Fen Shui (Dividing water): Drains dampness, stimulates urine discharge and relieves abdominal distension
4 Hoof	Ti Tou (Hoof): Relieves abdominal pain, indigestion and depressed appetite
5 Between Tail Root and Anus	Hou Hai (Caudal sea): Relieves paralysed rectum

Fig. 5.15 Acupuncture points for treating obstructive colic.

□ Herbal treatment. Dang Gui Cong Rong (*Angelica acutiloba* and *Cistanche deserticola*) Decoction

Ingredients and action:
Angelica acutiloba (Dang Gui) 200 g
(first stir-fried with oil)
Nourishes and invigorates blood, moisturizes intestines
Cistanche deserticola (Rou Cong Rong) 100 g
(first stir-fried with wine)
Loosens intestines to relieve constipation
Cassia angustifolia Vahl (Fan Xie Ye) 60 g
Drains heat and eliminates stagnation
Saussurea lappa (Guang Mu Xiang) 15 g
Relieves stagnation and promotes qi flow
Magnolia biondii (Hou Po) 30 g
Relieves stagnation and promotes qi flow
Citrus aurantium (Zhi Ke, mature fruit, stir-fried) 30 g
Relieves stagnation and promotes qi flow
Cyperus rotundus (Xiang Fu) 30 g
(first stir-fried with vinegar)
Relieves stagnation and promotes qi flow
Dianthus superbus (Qu Mai) 15 g
Stimulates urine discharge and clears heat
Tetrapanax papyriferus (Tong Cao) 10 g
Stimulates urine discharge and drains damp heat
Massa fermentata medicinalis (Liu Qu) 60 g
Relieves stagnation and promotes qi flow

Sesame oil 200–250 ml
Nourishes gastrointestinal tract and reinforces efficacy of other herbs

The ingredients are simmered together to make a decoction.

□ Acupuncture. As the above chart (Fig. 5.15) shows, the acupuncture points used in treating obstructive colic differ from those for spasmodic colic as a major purpose of the treatment is to remove the obstruction, which often occurs in the large intestine.

□ Manual removal of faeces. The clinician inserts a hand well into the rectum to squash or break up any impacted faeces or to remove solids from the body to eliminate the obstruction (Fig. 5.16).

□ Supplementary measures. Repeated lying down and rising up as well as rolling can cause the horse to stumble or a displacement of the stomach from its correct position. It is important to get a colicky horse on its feet to prevent injuries. Walking the horse to stimulate digestion and blood circulation can also help.

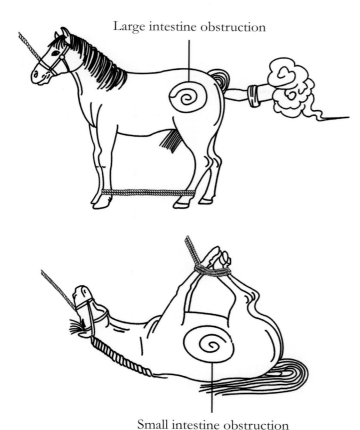

Large intestine obstruction

Small intestine obstruction

Fig. 5.16. Rectal exploration and manual removal of faeces. This late Ming dynasty illustration shows procedures employed for obstructive problems of the gastrointestinal tract. (Reference: *Yuan Heng's Collection of Horse Treatments*)

Once the intestinal blockage is resolved, the animal should fast for one or two meals and then gradually return to a normal diet.

Proper husbandry and dietary management are essential in the prevention of impaction. The horse needs a constant supply of clean water and a hygienic living environment. Avoid too much or too little exercise.

As the pain caused by obstructive colic can be immediate and intense, a combination of Western medication and TCVM is usually advisable.

GASTROENTERITIS

Gastroenteritis is inflammation of the stomach, and small and large intestine, and usually results in loose stools or diarrhoea. Gastroenteritis can be acute or chronic.

AETIOLOGY AND PATHOGENESIS

Overexertion in extreme summer heat resulting in penetration of heat and damp pathogens, overfeeding a starved horse or poor quality of feeds and dirty drinking water resulting in lodging of heat pathogens in a horse's visceral organs all lead to disorders of qi movement. The intestines become unable to digest and absorb food and gastroenteritis develops.

CLINICAL SIGNS OF ACUTE GASTROENTERITIS

Acute gastroenteritis can result in fever and lethargy, loss of appetite, diarrhoea, fetid and even bloodstained stools, preference for cold water, a dark-red or yellowish oral cavity, foul breath, and a surging and rapid pulse.

TREATMENT

The treatment of acute gastroenteritis aims at dispelling heat pathogens, detoxification, and relief of inflammation and pain.

▫ Herbal treatment. Yu Jin (*Curcuma aromatica*) Powder I

Ingredients and action:
Curcuma aromatica (Yu Jin) 45 g
  Cools blood, resolves stagnation, promotes qi flow
Terminalia chebula (He Zi) 30 g
  Intestinal astringent stops diarrhoea
Scutellaria baicalensis (Huang Qin) 30 g
  Clears damp-heat in intestines
Rheum officinale (Da Huang) 45 g
  Purges fire, clears heat, invigorates blood
Coptis chinensis (Huang Lian) 30 g
  Drains dampness and clears heat in intestines
Gardenia jasminoides (Zhi Zi) 30 g
  Cools blood and stops bleeding
Paeonia lactiflora (Bai Shao) 20 g
  Nourishes blood and relieves pain
Phellodendron amurense (Huang Bai) 30 g
  Clears heat, eliminates dampness, drains fire, releases toxins

The herbs are ground into a fine powder and mixed with water before administering.

The prescription is based on a classical recipe from *Yuan Heng's Collection of Horse Treatments*.

Chronic gastroenteritis can often develop from inadequately treated acute gastroenteritis, but it can also occur without a preceding acute episode. It presents with low spirits, a dull coat, a depressed appetite, loose faeces, slight abdominal pain, and dull and thinning hair. Severe cases result in repeated lying down and getting up, lying down and looking at the stomach (Fig. 5.17), and a soft, formless, dark-maroon, fetid stool.

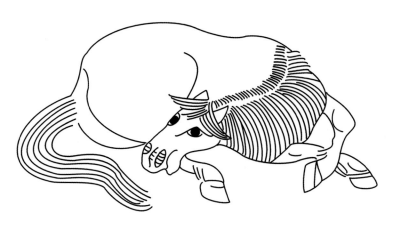

Fig. 5.17 A sign of chronic gastroenteritis: lying down and looking at the stomach. (Reference: *Simu's Collection of Equine Medicine*)

TREATMENT

Treatment involves dispelling heat pathogens, detoxification, promotion of qi movement and tonification of the spleen.

▫ Herbal treatment. Yu Jin (*Curcuma aromatica*) Powder II

Ingredients and action:
Curcuma aromatica (Yu Jin) 75 g
　Cools blood, resolves stagnation, promotes qi flow
Rheum officinale (Da Huang) 30 g
　Purges fire, clears heat, invigorates blood
Coptis chinensis (Huang Lian) 30 g
　Drains fire and dampness, relieves toxins, clears heat
Scutellaria baicalensis (Huang Qin) 30 g
　Clears heat, dries dampness, reduces fire, eliminates toxicity
Glycyrrhiza uralensis (Gan Cao) 15 g
　Tonifies spleen, promotes qi flow, relieves inflammation and pain, mediates properties of other herbs
Angelica acutiloba (Dang Gui) 30 g
　Invigorates blood and smoothes qi flow
Carthamus tinctorius L. (Hong Hua) 18 g
　Improves blood circulation, dredges meridians, dispels stasis, alleviates pain
Rhizoma anemarrhenae (Zhi Mu) 30 g
　Clears heat, purges fire, nourishes Yin, moistens dryness
Radix trichosanthis (Tian Hua Fen) 30 g
　Clears heat, purges fire, generates body fluids, quenches thirst, moistens dryness
Angelica dahurica (Bai Zhi) 12 g
　Dispels cold and relieves pain

The ingredients are ground into a fine powder and mixed with water before being administered.

　The prescription is based on *Qinghai TCVM Case Study Collection*.

▫ Supplementary measures. If a sick horse has lost its appetite for more than four to five days, force-feed with porridge, or consider giving a mix of salt, water and glucose. When recovery starts, the animal should be fed with an easily digestible food. Proper dietary and husbandry practices should be instituted and maintained long term.

DISCLAIMER

All the materials presented in this chapter are meant to serve as a guide for informational purposes only and should not take the place of a consultation with a trained TCVM clinician.

Photo Credits

The author acknowledges the courtesy given to use the images in this book. Their copyright remains with the following providers:

A Rui/Microfotos: Fig. 4.10

Arthorse/stock.adobe.com: Fig. 4.17

Caoyuan Ru Feng/Microfotos: Figs. 1.18-1.19

Chuntian de Laojiu/Microfotos: Fig. 4.6

Dong Li/Microfotos: Fig. 2.28

Emperor Qin Shi Huang's Mausoleum Site Museum: Figs. 2.1-2.2, 4.5

Gansu Provincial Museum: Fig. 2.6

Hong Kong Racing Museum: Figs. 3.2, 4.20

Huitu: Figs. 2.27, 3.9, 4.11, 4.14-4.15

Iwan Baan: Figs. 1.16-1.17

Kerri-Jo Stewart: Fig. 4.16

Li Donghui/Microfotos: Fig. 3.10

Li Jianpeng/Microfotos: Fig. 1.14

Li Xiaoming/Microfotos: Fig. 1.12

Liu Zhaoming/Microfotos: Fig. 1.15

Long Jiaze/Microfotos: Figs. 1.1, 1.4

Maoling Museum: Figs. 2.3-2.5

Marilyna/Microfotos: Fig. 5.11

Master2/Microfotos: Fig. 4.19

Metropolitan Museum of Art, New York, Purchase, Friends of Asian Art Gifts, 1991: Figs. 2.24-2.26

Metropolitan Museum of Art, New York, Purchase, The Dillon Fund Gift, 1977: Fig. 2.14

National Museum of China: Figs. 2.16-2.18, 3.12

National Palace Museum: Fig. 2.23

North Wind Picture Archives/Alamy Stock Photo: Fig. 4.13

Osaka City Museum of Fine Arts: Fig. 2.20

Panoramic Vision: pp. 4-5, Figs. 2.7-2.9, 2.12, 3.7-3.8, 4.12

Penn Museum, image #C395: Fig. 2.10

Penn Museum, image #C396: Fig. 2.11

Qin Jiaqiang/Microfotos: Fig. 1.11

River Mountain/Microfotos: Fig. 4.2

Shaanxi History Museum: Fig. 2.15

Shi Gangang/Microfotos: Fig. 3.11

Sun Zhiyuan/Palace Museum: Fig. 2.13

Thierry/stock.adobe.com: Fig. 4.7

TNM Image Archives: Fig. 2.19

Vikarus/Microfotos: Fig. 4.18

Wang Jinhui/Palace Museum: Fig. 2.22

Wang Zhanhong/CPC Liaoning Animal Domestic Genetic Resources Conservation and Utility Centre: Fig. 4.22

Wu Mingcheng and Xiang Mingyi/Heilongjiang Animal Reproduction Supervising Station and Northeast Agricultural University: Fig. 4.21

Xinhua: Figs. 3.3-3.5, 4.1, 4.3-4.4, 4.8-4.9

Xu Beihong Memorial Museum: Figs. 2.27-2.28

Yan Hong/Microfotos: Fig. 3.1

Ye Hongling/Microfotos: Fig. 1.5

Young Yin Hung: Figs. 1.2-1.3, 1.6-1.10, 1.20, 5.1-5.7, p. 161

Yu Ningchuan/Palace Museum: Figs. 1.13, 2.21

Zhou Shihua/Microfotos: Fig. 3.6

Ziranyuyanlu/Microfotos: pp. 2-3, 6

BIBLIOGRAPHY

BOOKS

Barnhart, R. M., Yang, X., Nie, C. Z., Cahill, J., Wu, H., Lang, S. J. and Peck, J. (2002). *Three Thousand Years of Chinese Painting*. New Haven, CT: Yale University Press.

Blood, D. C., Henderson, J. A. and Radostits, O. M. (1979). *Veterinary Medicine: A Textbook of the Diseases of Cattle, Sheep, Pigs, and Horses*. Philadelphia: Lea & Febiger.

Bowling, A. T. and Ruvinsky, A. (2000). *The Genetics of the Horse*. Wallingford, Oxon: CABI.

Chen, C. X. (2003). *Xu Beihong*. Shijiazhuang: Hebei Education Press.

Chen, S. F. (2009). *History of the Discovery of Petroglyphs in China*. Shanghai: Shanghai People's Press.

Ciarla, R. (Author) and Luca, A. D. (Photographer). (2012). *The Eternal Army: The Terracotta Soldiers of the First Emperor*. Auckland: White Star Publishers.

Coates, A. (1984). *China Races*. Hong Kong: Oxford University Press (China) Ltd.

Cui, L. C. (1998). *Sports in Ancient China*. Beijing: China Encyclopedia Press.

Cui, L. C. (2000). *Album of Ancient Sports Art in China*. Beijing: Chung Hwa Book Company.

Dossenbach, M. and D. (1994). *The Noble Horse*. Auckland: David Bateman Limited.

Fan, K. (2005). *Differentiation, Application and Analysis of Formulas of Traditional Chinese Veterinary Medicine*. Beijing: Chemical Industry Press.

Flaws, B. (1997). *The Secret of Chinese Pulse Diagnosis*. Colorado: Blue Poppy Enterprises, Inc.

He, L. C. (2006). *Illustrated Qin Shi Huang Mausoleum*. Chongqing: Chongqing Publisher.

Hendricks, B. L. (2007). *International Encyclopedia of Horse Breeds*. Norman: University of Oklahoma Press.

Hong Kong Museum of Art and Xu Beihong Museum. (1988). *The Art of Xu Beihong*. Hong Kong: Hong Kong Urban Council.

Hu, J. and Zhong, X. Z. (2007). *Xinjiang Password*. Beijing: Five-Continent Communications.

Hu, X. L. (2002). *Tang Sancai: Examination and Viewing*. Nanchang: Jiangxi Arts Press.

Inner Mongolia Autonomous Region Bureau of Statistics. *Inner Mongolia Statistical Yearbooks (2010-2016)*. Beijing: China Statistics Press.

Jiang, C. S. (1991). *A Chinese-English Dictionary of Traditional Chinese Veterinary Medicine*. Beijing: China Agricultural University Press.

Jiang, C. W. and Chen, Y. F. (2008). *Traditional Chinese Veterinary Medicine*. Beijing: China Agricultural Press.

Jiang, H. Q. (2008). *Xu Beihong's Neighing Horses and Sad Songs*. Beijing: Oriental Press.

Johnson, R. L. (2003). *Ernest Shackleton: Gripped by the Antarctic*. Minneapolis: Carolrhoda Books.

Kelekna, P. (2009). *The Horse in Human History*. Cambridge: Cambridge University Press.

Kentucky Horse Park. (2000). *Imperial China: The Art of the Horse in Chinese History*. Chicago: Art Media Resources Ltd.

Klide, A. M. and Kung, S. H. (1977). *Veterinary Acupuncture*. Philadelphia: University of Pennsylvania Press.

Klimuk, A. (Author) and Baboev, A. (Photographer). (2014). *Golden Horse: The Legendary Akhal-Teke*. New York: Harry N. Abrams.

Kong, Y. C. (2005). *The Cultural Fabric of Chinese Medicine*. Hong Kong: Commercial Press.

Langdon, J. (1986). *Horses, Oxen and Technological Innovation*. Cambridge: Cambridge University Press.

Li, L. X., He, L., Liu, Z. and Yuan, H. M. (2010). *Arts of China: Hardware*. Nanjing: Nanjing University Press.

Li, S. (1957). *Simu's Collection of Equine Medicine*. Shanghai: Chung Hwa Book Company.

Liu, K. Z. (2008). *Han Bricks with Colour Decoration*. Beijing: Party School of the Central Committee of C.P.C.

Liu, S. B., Du, Y. C. and Han, G. C. (Eds.). (2005). *China Horse Industry Thesis Collection*. Beijing: China Agricultural Science and Technology Press.

Liu, Y. H. (2013). *Carriages and Harness of Ancient China*. Beijing: Tsinghua University Press.

Lu, H. C. (1994). *Chinese System of Natural Cures*. New York: Sterling Publishing Co., Inc.

Mak, P. and Bazelaire, M. D. (1997). *Heavenly Horses/Le Cheval Chinois*. Paris: Hermes.

McGorum, B. C., Dixon, P. M., Robinson, N. E. and Schumacher, J. (Eds.). (2006). *Equine Respiratory Medicine and Surgery*. Philadelphia: Elsevier Health Science.

National Bureau of Statistics of China. *China Statistical Yearbooks* (2000-2018). Beijing: China Statistics Press.

Needham, J., Wang, L., Yates, R. D. S., Lu, G. D. and Ho, P. Y. (1987). *Science and Civilisation in China: Volume 5, Chemistry and Chemical Technology.* Cambridge: Cambridge University Press.

Pan, J. W. (2008). *Home Town of the Heavenly Horse: Yinwuwei.* Lanzhou: Gansu Culture Press.

Pickeral, T. (2009). *The Horse: 30,000 Years of the Horse in Art.* London: Merrell Publishers Limited.

Portal, J. (2007). *The First Emperor: China's Terracotta Army.* Cambridge: Harvard University Press.

Radostits, M., Gay, C., Hinchcliff, K. and Constable, P. (Eds.). (2007). *Veterinary Medicine: A Textbook of the Diseases of Cattle, Sheep, Pigs, Goats and Horses.* Philadelphia: Elsevier Health Science.

Riordan, J. and Jones, R. (1999). *Sport and Physical Education in China.* London and New York: E. & F. N. Spon.

Sullivan, M. (1984). *The Arts of China.* Berkeley: University of California Press.

The Hong Kong Museum of History. (2012). *The Majesty of All Under Heaven: The Eternal Realm of China's First Emperor.* Hong Kong: The Hong Kong Museum of History.

Wang, D. D. (2012). *Ancient Ceramic Examination: A Beginner's Guide.* Hefei: Anhui Science and Technology Press.

Wang, D. F. and Xu, X. L. (2006). Inner Mongolia: *The Horseback People.* Beijing: Foreign Press.

Xie, H. S. and Preast, V. (Eds). (2010). *Xie's Chinese Veterinary Herbology.* Ames, IA: John Wiley and Sons.

Xie, H. S. and Preast, V. (Eds). (2007). *Xie's Veterinary Acupuncture.* Ames, IA: Blackwell Publishing.

Yang, H., Hang, K. and Zheng, Y. (2008). *Chinese Equine History.* Hong Kong: Commercial Press.

Yu, B. Y. and Yu, B. H. (1957). *Yuan Heng's Collection of Horse Treatments.* Beijing: Chung Hwa Book Company.

Yu, C. and Zhang, K. J. (Eds.). (2003). *Classical Archives of Traditional Chinese Veterinary Medicine.* Beijing: China Agricultural University Press.

Yu, H. (2014). *2000 Years of Equestrian Painting in China.* Shanghai: Shanghai Shuhua Publisher.

Yuan, Z. Y. (2012). *The Eternal Legion of China's First Emperor: The Exploration of Qin's Terracotta Army.* Hong Kong: Hong Kong Open Page.

Zhang, K. (2011). *History of Cavalry in China.* Beijing: Chinese Liberation Army Press.

Zheng, P. L., Zhang, Z. G., Chen, X. H. and Tu, Y. Y. (Eds.). (1987). *Equidae Breeds Catalog in China.* Shanghai: Shanghai Scientific & Technical Publishers.

REPORTS

(2003, June). *Report on Domestic Animal Genetic Resources in China.* Beijing: Food and Agriculture Organization of the United Nations.

(2007). *List of Breeds Documented in the Global Databank for Animal Genetic Resources.* Rome: Food and Agriculture Organization of the United Nations.

(2007). *WHO International Standard Terminologies on Traditional Medicine in the Western Pacific Region.* New York: WHO.

(2009). *Chinese Equine Industry Overview.* Beijing: Kentucky China Trade Centre.

(2009). *Sustainable Agriculture for People and the Planet: Animal Welfare in Farming.* London: World Society for the Protection of Animals.

(2011, May). *Gain Report: Horse Market Profile.* Washington, DC: USDA Foreign Agricultural Service.

Wang, J. M. (2011). *Impacts of Enforcing Grazing Ban on Production and Lives of Herders.* Beijing: Chinese Academy of Agricultural Sciences.

JOURNALS

(2015, October 19). Over 70% of the 500 Riding Clubs in China Cannot Make Money. *Oriental Outlook.*

Chen, X. (2006). The Cultural Connotations of the Horse Image in Chinese Painting and its Artistic Achievements. *Art Panorama, Volume 6.*

Davis, B. (2007). Timeline of the Development of the Horse. *Sino-Platonic Papers,* 177. Retrieved from http://sino-platonic.org/complete/spp177_horses.html

Guo, K. X. (2002). Ancient Equestrian Culture on Two Select Areas. *Agricultural Archaeology, Volume 3,* 271-280.

Guo, W. C. (2006). The Role of the Horse in Chinese History. *Shantou University Journal (Humanities & Social Science Bimonthly), Volume 22 (1),* 15-18.

Guo, W. (2008). Comparison of the Horse's Artistic Image in the Song of Odes and the Zhuangzi. *Modern Chinese, Volume 10,* 22-23.

Hong, L. (2010). Discussion of the Horse's Image in the Song of Odes. *New West, Volume 10,* 97-98.

Huang, C. Z. (1994). Horse Rearing Business in Ancient Liangshan. *Agricultural Archaeology, Volume 1,* 310-314.

Jie, D. W. (1995). The Yunnan Pony Research. *The Journal of Animal Husbandry and Veterinary Medicine in Yunnan, S1*, 7-10.

Jing, C. (2002). Blood Sweating Purebreds Flying from the West Region. *Spring and Autumn of Literature and History, Volume 12*, 42-44.

Li, C. (1996). Horse Rearing Practices in Ancient China. *Ancient and Modern Agriculture, Volume 3*, 14-23.

Li, C. G. (2008). The Change of Equestrian Art Style from the Han Dynasty to the Tang Dynasty. *Art Observation, Volume 8*, 111.

Lin, L. (2000). The Origin of Polo. *Journal of Guizhou Literature and History, Volume 6*, 9-13.

Lin, Y. F. and Li, J. (2003). The Yili Horse under Wusun Mountains. *Encyclopedia Knowledge, Volume 12*, 54-55.

Liu, X. S. (2004). Art Technique of "Shepherd Horse". *Journal of Yuncheng University, Volume 22 (4)*, 71-73.

Liu, Y. L. (2005). Dancing Horse and Horse Dancing. *Chinese Culture Research, Volume 3*.

Muhetaer, Wang, R., Shi J., Shalitana and Ayiding. (2000). Relationship between the Xinjiang Horse's Geographical Distribution (with Precipitation and Temperature Variation) and its Characteristics. *Pratacultural Science, Volume 9 (6)*, 64-71.

The Chinese National Committee for Man and the Biosphere. (2010). *Mongolian Horse. Man and the Biosphere*.

The Chinese National Committee for Man and the Biosphere. (2013). *The Przewalski's Horse. Man and the Biosphere, Volume 5*.

Wang, H. S. (2006). The Role of the Horse in the Han Dynasty and the Equine Sculpture Accomplishment. *Beauty and Times, Volume 5*, 36-37.

Wang, L. (2017). Contexts and Media: Evolution of the Image of "Six Steeds of Zhaoling Mausoleum" from the Tang to the Song and Jin Dynasties. *Northern Art, Volume 3*, 90-95.

Wei, S. C. (1997). Traditional Chinese Veterinary Medicine before the Qin Dynasty. *Agricultural Archaeology, Volume 3*, 268-269.

Xiang, C. T. (2005). Phenomenological Theory of Traditional Chinese Veterinary Medicine. *The Journal of Animal Husbandry and Veterinary Medicine in Sichuan, Volume 8*.

Xu, C. T. and Yi, W. (1999). From Painted Stone and Brick Friezes to Bronze Sculpture. *Jianghan Archaeology, Volume 3*, 65-67.

Xu, Q. L. (2009). Interpretation of the Horse in the Yi-Jing. *Journal of Hunan Institute of Humanities, Science and Technology, Volume 6 (12)*, 94-96.

Yan, X. and Cui, T. B. (2002). Relationship between the Hequ's Ecological Characteristics and its Habitat. *Pratacultural Science, Volume 19 (6)*, 71-74.

Yao, Y. B., Zhang, X. Y. and Dong, J. H. (2004). Ecological and Climatic Conditions for the Hequ Horse. *Chinese Journal of Animal Science, 40 (2)*, 30-33.

Zhang, Y. J. and Wu, X. Z. (2002). The Artistic Image and Presentation Techniques of the Horse in the Qin and Han Dynasties. *Southeast Culture, Volume 5*, 48-53.

Zhu, F. P. (2006). The Horse as a Porcelain Motif. *Jingde Town Porcelain, Volume 16 (3)*, 5-6.

ESSAY/THESIS

Tharsen, J. R. (2006). *Sport and the "Competitive Spirit" in Ancient China*. Retrieved from http://home.uchicago.edu/~jcarlsen/2007/downloads/SportAndCompetitionInAncientChina_eng.pdf

Wang, S. (2004). *Burden on the Horse Back: Discussion of Chinese and Western Equine Sculptures Borne with or without Riders*. (Unpublished master's dissertation). Nanjing Art College.

WEBSITES

621m.cn. *Part Five: The Five Elements Theory – Natural Philosophy in Ancient China*. Retrieved 2012, February 4, from http://www.621m.cn/search/doc/50053

Access Tibet Tour. *Horse Racing Fair and Archery Festival*. Retrieved 2012, February 6, from http://www.accesstibettour.com/horse-racing-festival.html

AGTech. *China: A Rapidly Growing, Under-Penetrated Lottery and Wagering Market*. Retrieved 2017, February 27, from http://www.agtech.com/html/industry_lottery_overview_char.php

Alternative Veterinary Medicine Centre. Retrieved 2007, May 4, from http://www.alternativevet.org/articles_header.htm#horses

Animal Welfare Education. *Why Is Animal Welfare Education Important?* Retrieved 2011, February 9, from http://www.animalwelfare-education.eu/information/state-of-play.html

Asian Racing. *Beijing Jockey Club Shut Down*. Retrieved 2012, February 8, from http://www.asianracing.nu/vb/showthread.php?t=2279

A World of Chinese Medicine. Retrieved 2012, February 4, from http://www.aworldofchinesemedicine.com

Baidu. *Akhal-Teke*. Retrieved 2012, January 30, from http://baike.baidu.com/view/27748.htm

Baidu. *Humane Slaughter*. Retrieved 2012, February 8, from https://baike.baidu.com/item/人道屠宰/9956715?fr=aladdin

Baidu. *Six Steeds of the Zhaoling Mausoleum*. Retrieved 2012, January 29, from http://baike.baidu.com/view/2435.htm

Baidu. *The Guoxia Horse*. Retrieved 2012, January 30, from http://baike.baidu.com/view/358381.htm

Baidu. *The Przewalski's Horse*. Retrieved 2012, January 30, from http://baike.baidu.com/view/81510.htm#8

Baidu. *The Yili Horse*. Retrieved 2012, January 29, from http://baike.baidu.com/view/117378.htm

Baike. *Red Hare*. Retrieved 2017, January 30, from http://baike.baidu.com/view/43462.htm

Baike. *Silver Saddle-flask with Gilded Dancing Horses*. Retrieved 2017, January 29, from http://www.baike.com/wiki/鎏金舞马衔杯纹银壶&prd=so_1_doc

Baike. *The Przewalski's Hors*e. Retrieved 2017, January 30, from http://www.baike.com/wiki/普氏野马&prd=so_1_doc

Bayaer. Chinahorse.org. *Worship of the Horse God at the Genghis Khan Mausoleum*. Retrieved 2007, March 23, from http://www.chinahorse.org/html/74.html

Beijing Palace Museum. *The Fat and Lean Horses*. Retrieved 2012, February 8, from http://www.dpm.org.cn/www_oldweb/Big5/phoweb/Relicpage/1/R230.htm

Björkell, S. *All Bets are On! Lottery Games and Gambling in China*. Retrieved 2008, July 30, from http://radio86.com/focus/all-bets-are-lottery-games-and-gambling-china

Chen, X. and Huang, S. (2007, December 28). China.org.cn. *Przewalski's Horse Project Short of Fund*s. Retrieved 2010, December 12, from http://www.china.org.cn/english/environment/237442.htm

China (Shanghai) International Horse Fair. *2010 Review*. Retrieved 2012, February 8, from http://www.horfachina.com/web_en/Show_news.asp?C_ID=5&Art_id=1147

China.com. *Horse Policy and Livestock Industry in the Yuan Dynasty*. Retrieved 2005, May 18, from http://military.china.com/zh_cn/history2/06/11027560/20050518/12324702_1.html

China.org.cn. *Horse Racing Trial Starts in Wuhan*. Retrieved 2012, February 1, from http://www.china.org.cn/2008-04/05/content_14336017.htm

China.org.cn. *Humane Slaughtering Spearheads China's Drive to Promote Animal Welfare*. Retrieved 2009, September 5, from http://www.china.org.cn/environment/2009-09/05/content_18519951.htm

China.org.cn. *Proposed Animal Welfare Law Watered Down*. Retrieved 2010, January 26, from http://www.china.org.cn/china/2010-01/26/content_19309286.htm

China.org.cn. *Tomb of General Huo Qubing of the Han Dynasty*. Retrieved 2012, January 29, from http://www.china.org.cn/english/features/atam/115101.htm

China.org.cn. *Xu Beihong Museum*. Retrieved 2012, January 29, from http://www.china.org.cn/english/TR-e/41443.htm

China.org.cn. *Zhaoling Mausoleum of the Tang Dynasty (618-907)*. Retrieved 2012, February 5, from http://www.china.org.cn/english/features/atam/115391.htm

Chinacultural.org. *Chinese Famous Horses*. Retrieved 2012, January 29, from http://www.chinaculture.org/library/2008-01/07/content_21049.htm

Chinaculture.org. *Horse Stepping on a Xiongnu Soldier and Horse Stepping on a Swallow*. Retrieved 2012, January 29, from http://www.chinaculture.org/gb/en_artqa/2005-12/28/content_77549_3.htm

Chinahorsefair.com. *Interview with Chinese Equestrian Association*. Retrieved 2010, June 21, from http://www.chinahorsefair.com.cn/

Chinaknowledge.de. *Chinese History – Zhou Dynasty Government and Administration*. Retrieved 2012, February 25, from http://www.chinaknowledge.de/History/Zhou/zhou-admin.html

China National Horse Industry Web. *The Yili Horse*. Retrieved 2012, February 4, from http://www.chinahorse.org/html/160.html

China Wildlife Conservation Association. *The Death of Przewalski's Horses*. Retrieved 2012, January 30, from http://zdx.forestry.gov.cn/portal/bhxh/s/713/content-92262.html

CINtcm.com. *Revelation of the Mystery of the Meridian*. Retrieved 2012, February 4, from http://www.cintcm.com/e_cintcm/e_forum/rev_m.htm

Confucius Institute Online. *Theory of Yin and Yang*. Retrieved 2009, August 24, from http://www.chinese.cn/tcm/en/article/2009-08/24/content_10169.htm

Cultural China. *Six Steeds of the Zhaoling Mausoleum*. Retrieved 2012, January 29, from http://arts.cultural-china.com/en/

Cultural China. *The Silver Pot with Dancing Horse Patterns*. Retrieved 2012, February 7, from http://history.cultural-china.com/en/54History869.html

Damo-qigong.net. *Qi Theory and Pathogenic Factors*. Retrieved 2012, February 4, from http://www.damo-qigong.net/pathogen1.htm

Davis, B. (2012, August 21). Horse of the Americas. *The Politics of Color – Appaloosa*. Retrieved 2019, November 13, from http://horseoftheamericas.com/colorpolitics.htm

Day, C. (1996, May). Alternative Veterinary Medicine Centre. *Herbal Medicine for Horses*. Retrieved 2006, February 7, from http://www.alternativevet.org/WS109-07.pdf

Discovery TCM. Retrieved 2012, February 4, from http://tcmdiscovery.com/

Drben.net. *China Report: Mongol (Mongolian) Ethnic Minority in China*. Retrieved 2012, February 3, from http://www.drben.net/ChinaReport/Sources/Art_Arts_Culture/Ethnic_Minorities_in_China/Mongolian/Mongolians-Mongol-Ethnic_Minority.html

Dyske. (2009, June 5). *The Chinese Lacking in Creativity?* Retrieved 2010, July 6, from http://alllooksame.com/?p=241

Equinenaturaltherapy.com. *Equine Massage*. Retrieved 2012, February 4, from http://www.equinenaturaltherapy.com/equine_massage.htm

Equisearch.com. *Respiratory Noises in Horses*. Retrieved 2012, February 6, from http://www.equisearch.com/horses_care/health/illnesses_injuries/eqnoise738/

Food and Agriculture Organization of the United Nations. *Livestock Primary*. Retrieved 2019, May 13, from http://faostat3.fao.org/download/Q/QL/E

Food and Agriculture Organization of the United Nations. *The Domestic Animal Diversity Information System*. Retrieved 2017, February 27, from http://dad.fao.org/

Globalmeatnews.com. *China's First Welfare Code for Feed Lots, Slaughterhouses Due in June*. Retrieved 2017, February 27, from http://www.globalmeatnews.com/Safety-Legislation/China-s-first-welfare-code-for-feed-lots-slaughterhouses-due-in-June

Globalmeatnews. *ICCAW President Defends China's Stance on Pig Welfare*. Retrieved 2017, February 27, from http://www.globalmeatnews.com/Environment/ICCAW-president-defends-China-s-stance-on-pig-welfare

Global Public Square. *China Bans Ancient Dog-Eating Festival after Online Uproar*. Retrieved 2011, September 22, from http://globalpublicsquare.blogs.cnn.com/2011/09/22/china-bans-ancient-dog-eating-festival-online-uproar/

Godfrey, M. (2009). *China International Business. Ready to Jump the Hurdle*. Retrieved 2009, February 10, from http://www.cibmagazine.com.cn/html/Print/Show.asp?id=827&ready_to_jump_the_hurdle.html

Heilbrunn Timeline of Art History. *Giuseppe Castiglione (Lang Shining): One Hundred Horses*. Retrieved 2012, January 29, from http://www.metmuseum.org/toah/works-of-art/1991.134

Hillier, B. (2014, February 6). Thoroughbredracing.com. *Horse Racing in China: Real, Surreal, or Virtual?* Retrieved 2016, March 13, from https://www.thoroughbredracing.com/articles/horse-racing-china-real-surreal-or-virtual-pt-iv/

Horse-disease.com. *Spasmodic, or Cramp, Colic*. Retrieved 2012, February 6, from http://www.horse-diseases.com/spasmodicorcrampcolic.html

Institute for Traditional Medicine. *The Kidney Network and Mingmen: Views from the Past*. Retrieved 2012, February 4, from http://www.itmonline.org/5organs/kidney.htm

International Union for Conservation of Nature and Natural Resources. *Equus ferus ssp. Przewalskii*. Retrieved 2017, February 27, from http://www.iucnredlist.org/details/7961/0

Learn TCM. *Traditional Chinese Medicine Science*. Retrieved 2012, February 6, from http://www.learntcm.com/articles/part-fifteen-pathogenesis.html

Lehane, B. (2010). *Pigs Get Time Off to Improve Pork Quality*. Retrieved 2010, March 6, from http://www.weirdasianews.com/2010/03/06/pigs-time-improve-pork-quality/

Li, D. J. (2004, 16 November). CRIOnline. *The Przewalski's Horse Returned Home*. Retrieved 2008, March 13, from http://gb.cri.cn/3821/2004/11/16/301@361641_1.htm

Liu, L. W. (2008, November 26). Time. *Horse Betting Back in China*. Retrieved 2012, February 3, from http://china.blogs.time.com/2008/11/26/horse-betting-back-in-china/

Mark Godfrey Blog. *Horses*. Retrieved 2012, February 3, from http://horses.markgodfrey.eu/#home.2

Megan, S. (2010, June 3). Care2.com. *Activist Spotlight: Enforce the China Animal Protection Law*. Retrieved 2011, March 3, from http://www.care2.com/causes/activist-spotlight-enforce-the-china-animal-protection-law.html#ixzz1lo2Vk2n7

Military Museum of the Chinese People's Revolution. *Cavalry and Stirrups*. Retrieved 2016, March 14, from http://www.jb.mil.cn/jszt/gbbtg/201303/t20130314_14406.html

Ministry of Finance. *2018 Sales of National Lottery*. Retrieved 2019, May 13, from http://m.mof.gov.cn/czsj/201901/t20190125_3132678.htm

Mosquito. (2010, July 11). Ponybox.com. *Horses of China*. Retrieved 2011, March 4, from http://www.ponybox.com/news_details.php?title=Horses-of-China&id=1019

Moxley, M. (2011, January 19). Globalgeopolitics.net. *Food Worries Rise in China*. Retrieved 2012, February 5, from http://globalgeopolitics.net/wordpress/2011/01/19/food-worries-rise-in-china/

National Treasure File. *Silver Saddle-flask with Gilded Dancing Horses*. Retrieved 2011, January 25, from http://www.youtube.com/

Official Website of the Chinese Olympic Committee. (2003, November 16). *A Brief Introduction to Ancient Sports in China*. Retrieved 2008, March 7, from http://en.olympic.cn/sports_in_ancient_china/2003-11-16/11313.html

Oklahoma State University. *Akhal-Teke*. Retrieved 2012, February 3, from http://www.ansi.okstate.edu/breeds/horses/akhalteke

Olhgroup. *Orient Lucky Horse*. Retrieved 2012, February 3, from http://www.olhgroup.com/en/aboutus.asp

Orientations. *The Thousand Li Horses*. Retrieved 2012, February 3, from http://publications.kaleden.com/articles/487.html

O'Reilly, B. (2012). Horsetalk.co.nz. *Ponies at the Poles: a Proud History*. Retrieved 2013, September 19, from http://horsetalk.co.nz/2012/12/29/ponies-at-the-poles-proud-history/#ixzz2Z6Dyg9MG

Pitman, S. (2008, February). Naturalism and the Theory of Evolution. *Donkeys, Horses, Mules, and Evolution: The Phenotypic Effects of Chromosome Variation*. Retrieved 2012, March 7, from http://www.detectingdesign.com/donkeyshorsesmules.html

Racingmemories.hk. *Silkylight*. Retrieved 2013, September 19, from http://racingmemories.hk/?s=silkylight

Ralphmag.org. *Eating Your Transport*. Retrieved 2013, September 19, from http://www.ralphmag.org/HN/antarctic.html

Rootdown.us. *Acupuncture*. Retrieved 2012, February 6, from http://www.rootdown.us/Points/KI_18?query=shi+guan#Profile

Sacred Lotus. *Jing Luo (Channels and Collaterals | Meridians and Sub-Meridians)*. Retrieved 2012, February 4, from http://sacredlotus.com/acupuncture/channel_theory.cfm

Shen-nong.com. *Traditional Chinese Medicine*. Retrieved 2012, February 4, from http://www.shen-nong.com

Songer, M. (2009, August 19). Geoplace.com. *Return of the Yema: Geotechnology Helps Reintroduce Endangered Wild Horses*. Retrieved 2012, March 9, from http://www.geoplace.com

Sothebys. *A Very Rare Black and Sancai-Glazed Fergana Horse, Tang Dynasty, 8th Century*. Retrieved 2015, December 3, from http://www.sothebys.com/en/auctions/ecatalogue/2004/chinese-works-of-art-n07974/lot.596.html

Sun Bin Blog. (2005, June 3). *The Crippled Strategist*. Retrieved 2008, July 9, from http://sun-bin.blogspot.com/2005/06/sun-bin-crippled-strategist.html

Surag.net. *Grassland Salon: Attention to the Destiny of Mongolian Horses and Herders*. Retrieved 2010, November 9, from http://www.surag.net

Szczepanski, K. (2019, July 3). ThoughtCo. *The Invention of the Saddle Stirrup*. Retrieved 2019, October 1, from https://www.thoughtco.com/invention-of-the-stirrup-195161

TCM Basic. Retrieved 2012, February 4, from http://www.tcmbasics.com/

The Adventurists. Retrieved 2019, July 7, from https://www.theadventurists.com/adventures/mongol-derby/

The Foundation for the Preservation and Protection of the Przewalski's Horse. Retrieved 2012, January 29, from http://www.treemail.nl/takh/about/history.htm

The Long Horse Ride. *Shandan*. Retrieved 2012, February 8, from http://www.thelonghorseride.com/breeds.htm

The Three Kingdoms. *Chapter 25. Besieged in Tushan, Guan Yu makes three conditions; Relieved at Baima, Cao Cao beholds a marvel*. Retrieved 2012, January 29, from http://threekingdoms.com/025.htm

Thoroughbred Bloodlines. *Byerley Turk*. Retrieved 2012, February 8, from http://www.bloodlines.net/TB/Bios/ByerlyTurk.htm

Traditional Chinese Medicine Library. Retrieved 2012, February 4, from http://www.tcmlib.com/zy/html45/showdetail-343434362ce799bde88ab72c7a79.html

Traditional Chinese Medicine – World Foundation. Retrieved 2012, February 4, from http://www.tcmworld.org/what_is_tcm/

Troika. *Russian Extreme Breeds*. Retrieved 2012, January 29, from http://www.horses.ru/breeds/extreme.htm

Wikipedia. *Cruelty to Animals*. Retrieved 2012, February 8, from http://en.wikipedia.org/wiki/Cruelty_to_animals#China

Wikipedia. *Fergana Horse*. Retrieved 2012, January 29, from http://en.wikipedia.org/wiki/Ferghana_horse

Wikipedia. *Horse Colic*. Retrieved 2012, January 25, from http://en.wikipedia.org/wiki/Horse_colic

Wikipedia. *Mongol Derby*. Retrieved 2019, July 7, from https://en.wikipedia.org/wiki/Mongol_Derby

Wikipedia. *Yan Liben*. Retrieved 2012, January 29, from http://en.wikipedia.org/wiki/Yan_Liben

Winger, J. (2005). Smithsonian's National Zoo & Conservation Biology Institute. *A Spirited Return*. Retrieved 2012, January 29, from http://nationalzoo.si.edu/Publications/ZooGoer/2005/5/phorses.cfm

World Bank. *Project Profile: Restoring Grasslands and Improving Herders' Livelihood*. Retrieved 2011, August 2, from http://www.worldbank.org/en/news/2011/08/02/Restoring-grasslands-and-improving-herders-livelihood

Wowhorses.com. *Horse Colic*. Retrieved 2012, February 6, from http://www.wowhorses.com/horse-colic.html

XinhuaNet.com. *Looking for a More Remote Habitat for the Przewalski's Horse*. Retrieved 2012, January 30, from http://news.xinhuanet.com/newscenter/2009-03/14/content_11009506.htm

NEWS

(2007, April 11). China Imposes Grazing Ban to Restore Grasslands. *Reuters*. Retrieved from http://www.reuters.com/article/2007/04/11/idUSPEK287035

(2008, April 5). Horse Racing Trial Starts in Wuhan. *Shanghai Daily*.

(2008, December 1). Horse Racing Back on Wuhan Courses. *Xinhua News*.

(2008, December 11). History of Wuhan Horse Racing. *QQ.com*. Retrieved from http://news.qq.com/a/20081211/001685.htm

(2009, August 16). Discontent Villagers Blocked the Entrance to the Racing Track as the Village Committee Leased Out Their Land at Low Prices. *Jinghua News*.

(2009, November 5). Old Hankou Horse Racing Track was a Huge Casino. *Hubei Dagongbao*.

(2011, August 8). When Will the Horses Run? *People's Daily Online*. Retrieved from http://english.peopledaily.com.cn/102780/7562580.html

(2011, June 25). Tibetan Pastoralists Gave up Horses for Motorbikes. *Beijing News*. Retrieved from http://epaper.bjnews.com.cn

(2011, May 31). Inner Mongolia Establishes Grassland Ecological Compensation Mechanism. *People's Daily Online*. Retrieved from http://english.people.com.cn/90001/90776/90882/7396449.html

(2015, November 23). Humane and Sustainable Swine Farming Project Contract Signing Ceremony in Beijing. *Saier Livestock Farming Net*. Retrieved from http://chinaswine.org.cn/content-18-18908-1.html

(2015, October 29). Encouragement of Humane Slaughter. *Xinxi News*. Retrieved from http://epaper.xxsb.com/showNews/2015-10-29/271421.html

Lee, M. (Ed.). (2010, September 28). Gansu Tentatively Releases Przewalski's Horses into Wild. *People's Daily Online*.

Li, Y. S. (Ed.). (2019, July 5). Mating Season Coming Up for Przewalski's Horses. *Chinanews.com*. Retrieved from http://www.chinanews.com/sh/2019/05-07/8830245.shtml

Liu, D. L. (2019, April 10). Przewalski's Horses are Galloping Freely in Xinjiang Kalamaili Mountains. *Xinjiang Daily News*. Retrieved from http://xj.people.com.cn/n2/2019/0410/c186332-32828708.html

Nan, X. H. (2004, October 9). Przewalski's Horses already in the Wild for Three Years. *Southern Weekend*. Retrieved from https://tech.sina.com.cn/d/2004-10-09/1825437033.shtml

Sun, J. B. (Ed.). (2019, June 3). Results of Cooperation between Xinjiang and Gansu Breeding Centres: Five Przewalski's Foals Will Be Born This Year. *Chinanews.com*. Retrieved from http://www.chinanews.com/sh/2019/06-03/8854709.shtml

Wang, Z. Q. (2009, November 13). No Excuse for Starved Horses. *China Daily*.

Watts, J. (2010, January 26). Chinese Legal Experts Call for Ban on Eating Cats and Dogs. *The Guardian*. Retrieved 2011, November 9, from http://www.guardian.co.uk/environment/2010/jan/26/dog-meat-china

Wu, Q. (2009, January 22). Wuhan Horse Racing: Deregulation Trial. *Joint Publishing Lifestyle Weekly*.

Xiao Lin. (2019, April 3). Japanese Scholar Itakura Masaaki Talks About the Reappearance of the Original Five Tribute Horses and Grooms by Li Gonglin of the Song Dynasty. *The Paper*. Retrieved from https://www.thepaper.cn/newsDetail_forward_3239310

Index

The Unrivalled Horse:
A 4,000-Year History
in China

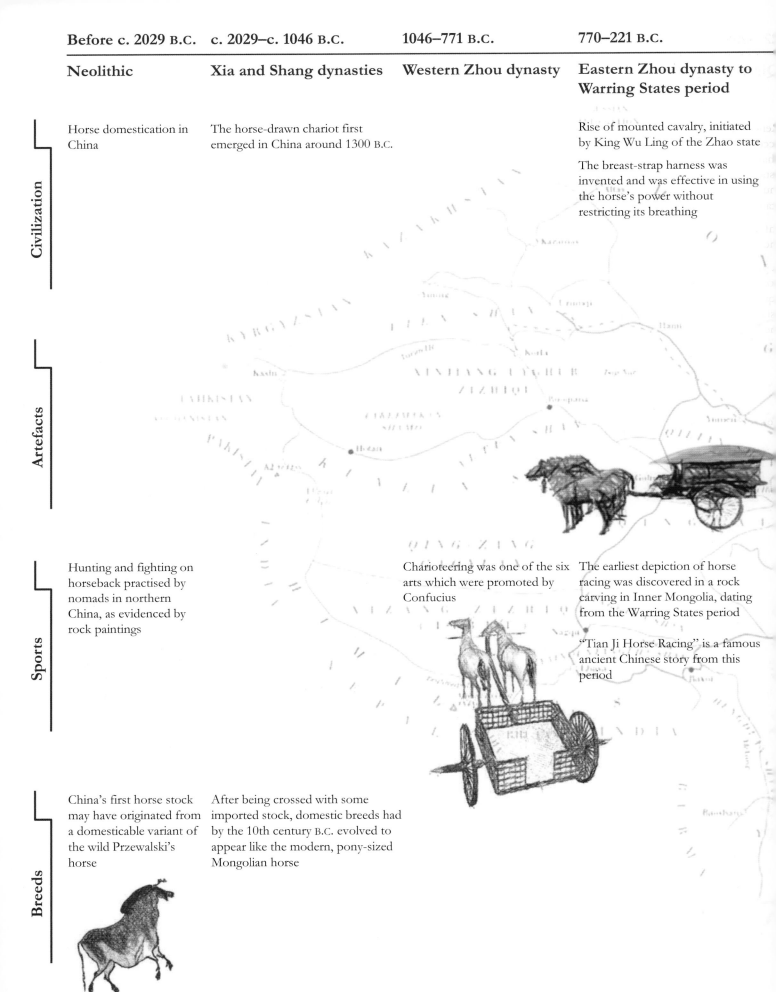

Before c. 2029 B.C.	c. 2029–c. 1046 B.C.	1046–771 B.C.	770–221 B.C.
Neolithic	**Xia and Shang dynasties**	**Western Zhou dynasty**	**Eastern Zhou dynasty to Warring States period**

Civilization

Horse domestication in China

The horse-drawn chariot first emerged in China around 1300 B.C.

Rise of mounted cavalry, initiated by King Wu Ling of the Zhao state

The breast-strap harness was invented and was effective in using the horse's power without restricting its breathing

Artefacts

Sports

Hunting and fighting on horseback practised by nomads in northern China, as evidenced by rock paintings

Charioteering was one of the six arts which were promoted by Confucius

The earliest depiction of horse racing was discovered in a rock carving in Inner Mongolia, dating from the Warring States period

"Tian Ji Horse Racing" is a famous ancient Chinese story from this period

Breeds

China's first horse stock may have originated from a domesticable variant of the wild Przewalski's horse

After being crossed with some imported stock, domestic breeds had by the 10th century B.C. evolved to appear like the modern, pony-sized Mongolian horse

in dynasty

Han dynasty

Three Kingdoms period to Southern and Northern dynasties

nowned for their highly
complished horsemanship and
arioteering skills, the Qin people
tablished the first unified,
ultinational and power-centralized
te in Chinese history

fantry rather than cavalry was
en the major military force

governing body in charge of
rses was set up to oversee
uestrian issues and its
tablishment continued throughout
e feudal period

rracotta horses

e terracotta horses
semble Hequ horses, a
tive breed

Invention of the horse collar (c. 100 B.C.) enhanced the horse's hauling power

The Han had already developed a mighty cavalry force strong enough to repel the increasing incursions of the Xiongnu

Han Wudi launched military quests against Fergana for "heavenly horses", bringing some 1,000 Nisaean/Akhal-Teke to China

Han Wudi sent Zhang Qian on a diplomatic mission to Wusun, which gave the Han scores of premium horses

Invention of the double-shaft carriage to increase horse availability

The saddle design was improved with a higher pommel and cantle, increasing the stability of the rider's seat and heralding the mass adoption of cavalry

Horse Stepping on a Xiongnu Soldier

Recumbent Horse

Prancing Horse

Bronze Cantering Horse

Horse dancing can be traced back to Emperor Han Wudi (reigned 140–87 B.C.) who received a Fergana horse trained to bow and move to the rhythm of drums

The Nisaean (blood-sweating purebred) and Akal-Teke arrived in China

Invention of stirrups in north-east China

Cavalry, especially heavy cavalry, marked an epoch

Polo first appeared in Chinese literature

618–907	960–1279	1279–1368	1368–1644
Tang dynasty	**Song dynasty**	**Yuan dynasty**	**Ming dynasty**

Civilization

Horse-breeding programmes culminated in the Tang dynasty, not only meeting the dynasty's appetite for horses but also improving the stock quality	The Song's major military power was infantry. The scale and strength of its cavalry were far lower than the Tang's and led to its overthrow by the Mongols	Punishments for horse-related crimes were harsh and cruel Focused on breeding Mongolian horses and decimated or ignored most other superior imported breeds, degrading the overall quality of horses in China A formidable and efficient mail system connected by horses was operated in the Mongol Empire	Significantly increased horse supply through for horses markets Gunpowder was widel used alongside infantry and cavalry

Artefacts

Six Steeds of Zhaoling Mausoleum (stone relief plaques) Night Shining White Silver Saddle-flask with Gilded Dancing Horses Tri-coloured glazed pottery horses	Five Tribute Horses and Grooms Six Steeds of Zhaoling Mausoleum (hand-scroll)	Emaciated Horse Mounted Official Fat and Lean Horses	

Sports

Horse dancing, polo and horseback hunting each reached an apex		The Mongols took over China, hosting the annual Naadam Festival	

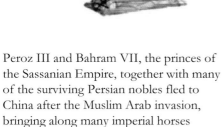

Breeds

Peroz III and Bahram VII, the princes of the Sassanian Empire, together with many of the surviving Persian nobles fled to China after the Muslim Arab invasion, bringing along many imperial horses Sino-Arab diplomatic relations formally started in 651 when several premium Arabian horses were presented to the Tang court	Nisaean horses were believed to have become extinct after the sack of Constantinople in the Fourth Crusade in 1204	Mongolian horses were the most common and treasured	The Admiral-eunuch Zheng He brought bac Arabian horses from h naval expeditions to Hormuz and East Afri

Qing dynasty   Republic of China   People's Republic of China

The Manchus subjugated the Ming
by their equestrian prowess

Towards the end of the dynasty,
foreign encroachment and gunboat
diplomacy prevailed. The cavalry of
the Manchus was virtually rendered
obsolete by the superior weaponry
and technology of the Western
powers

The Hundred Horses   Galloping Horse

A massive annual hunting event,
Mulanquimi, was held annually in the
early Qing

In 1798, the first modern horse race
in China was held in Macau, which
was a Portuguese possession

The British possession of Hong
Kong and other foreign concessions
in Tianjin, Shanghai, Beijing, Wuhan,
etc. staged race meetings even during
the wars

The Sanhe horse was very active in
racecourses across China and had
raced in Hong Kong from 1856
until 1972

In 1907, Sir Ernest Shackleton took
10 Sanhe horses on his first
expedition to Antarctica

The Bolshevik Revolution erupted
in 1917 causing tens of thousands
of Russians to flee into China
together with their horses, which in
time crossed with the native
Chinese stock, improving horse
breeds like the Sanhe and Yili

Automation is increasingly replacing horses, while the passion of
older nomads for the animals remains unabated

No horse rights and welfare laws in place and no progress in
legislation

The equestrian industry is not adequately developed. Instructors'
qualifications, horse transport, veterinarian standards, medical
facilities, insurance, to name but a few, all have ample room for
improvement

Horse racing completely disappeared in 1949 when the Communist
Party came to power

Betting on horse racing is illegal in China except in Hong Kong and
Macau

In the 1990s, China saw a rapid development of horse racing as
Nanjing, Wuhan, Beijing, Guangzhou, Dongguan, Shenzhen, Jinan
and Ningbo each had built a racecourse. However, most were either
shut down by the government under the "Strict Forbidden Order on
Horseracing Gambling" or finally switched to non-racing activities

Investors and local governments still live in hope that horse racing
will be liberated one day. They include global heavyweights Sheikh
Mohammed and also John Magnier's Coolmore, in partnership with
Teo Ah King's China Horse Club

The 2008 Olympic Equestrian Events were held in Hong Kong

Following the boom in China's economy in the 21st century,
equestrian pursuits have become a fashion statement among the
growing elite and expatriates

The Przewalski's horse was last seen in China in 1966 and was
reintroduced in 1985. Through the global endeavour of
conservationists, the rare species has in recent years been reclassified
from "extinct in the wild" to "endangered"

In tandem with China's modernization and urbanization, the value
of the horse has plunged. The numbers of major breeds like the
Sanhe and Hequ are decreasing while more and more other breeds
are at risk or endangered